To Pali bácsi,
der Zauberer von Budapest,
who got by with a little help from his friends
and achieved immortality with his proofs and conjectures


To Ron,
who made the last 39.76 percent of Paul's life easier
and gave generously of his time in helping me
understand Paul's world


To Ann,
who loves frogs, poetry, and me

# THE MAN WHO LOVED
# ONLY NUMBERS

# THE MAN WHO LOVED ONLY NUMBERS

## THE STORY

### OF PAUL ERDŐS AND

### THE SEARCH

### FOR MATHEMATICAL

### TRUTH

## BY PAUL HOFFMAN

HYPERION

NEW YORK

The author is grateful for permission from: THE CANADIAN BROADCASTING CORPORATION to
excerpt outtakes from the radio program "Math and Aftermath: A Tribute to Donald Coxeter,"
broadcast on the series *Ideas* on May 13, 1997. The program featured *Ideas* host Lister Sinclair and
was produced by Sara Wolch; RONALD GRAHAM, EDGAR PALMER, AND GORDON RAISBECK to
excerpt e-mail and snail mail to Ronald Graham; TOM TROTTER AND RICHARD GUY to excerpt
recollections of Paul Erdős that were posted on the Web; WILLIAM WEBB to excerpt Paul Erdős's
"Child Prodigies" (March 1971) in James Jordan and William A. Webb, editors, *Proceedings of the
Washington State University Conference on Number Theory* (Department of Mathematics and Pi
Mu Epsilon, Washington State University, Pullman, Washington); PARADE MAGAZINE and MAR-
ILYN VOS SAVANT to excerpt from "Ask Marilyn" columns of September 9, 1990, December 2,
1990, and February 17, 1991; THE CARTOON BANK to quote from a © 1998 cartoon by Robert
Mankoff, FABER AND FABER LTD to excerpt from *Arcadia* by Tom Stoppard. Author's failure to
obtain a necessary permission for the use of any other copyrighted material included in this work
is inadvertent and will be corrected in future printings of the work.

BOOK DESIGN BY DOROTHY S. BAKER

Library of Congress Cataloging-in-Publication Data
Hoffman, Paul, 1956–
The man who loved only numbers : the story of Paul Erdős
and the search for mathematic truth / Paul Hoffman.—1st ed.
p. cm.
Includes bibliographical references (p. 279) and index.
ISBN 0-7868-8406-1
1. Erdős, Paul, 1913– . 2. Mathematicians—Hungary—Biography.
I. Title.
QA29.E68H64 1998
510'.92—dc21 98-14027
[B] CIP

Trade Paperback ISBN: 978-0-7868-8406-3

FIRST PAPERBACK EDITION

11  13  15  17  19  20  18  16  14  12

# CONTENTS

———

# CONTENTS

Mathematical truth is immutable; it lies outside physical reality.... This is our belief; this is our core motivating force. Yet our attempts to describe this belief to our nonmathematical friends are akin to describing the Almighty to an atheist. Paul embodied this belief in mathematical truth. His enormous talents and energies were given entirely to the Temple of Mathematics. He harbored no doubts about the importance, the absoluteness, of his quest. To see his faith was to be given faith. The religious world might better have understood Paul's special personal qualities. We knew him as Uncle Paul.

—Joel Spencer

To find another life this century as intensely devoted to abstraction, one must reach back to Ludwig Wittgenstein (1889–1951), who stripped his life bare for philosophy. But whereas Wittgenstein discarded his family fortune as a form of self-torture, Mr. Erdős gave away most of the money he earned because he simply did not need it.... And where Wittgenstein was driven by near suicidal compulsions, Mr. Erdős simply constructed his life to extract the maximum amount of happiness.

—*The Economist*

# THE MAN WHO LOVED
# ONLY NUMBERS

# O

## THE TWO-AND-A-HALF-BILLION-YEAR-OLD MAN

Végre nem butulok tovább
(Finally I am becoming stupider no more)

 —the epitaph Paul Erdős wrote for himself

Paul Erdős was one of those very special geniuses, the kind who comes along only once in a very long while yet he chose, quite consciously I am sure, to share mathematics with mere mortals—like me. And for this, I will always be grateful to him. I will miss the times he prowled my hallways at 4:00 A.M. and came to my bed to ask whether my "brain is open." I will miss the problems and conjectures and the stimulating conversations about anything and everything. But most of all, I will just miss Paul, the human. I loved him dearly.

 —Tom Trotter

It was dinnertime in Greenbrook, New Jersey, on a cold spring day in 1987, and Paul Erdős, then seventy-four, had lost four mathematical colleagues, who were sitting fifty feet in front of him, sipping green tea. Squinting,

Erdős scanned the tables of the small Japanese restaurant, one arm held out to the side like a scarecrow's. He was angry with himself for letting his friends slip out of sight. His mistake was to pause at the coat check while they charged ahead. His arm was flapping wildly now, and he was coughing. "I don't understand why the SF has seen fit to send me a cold," he wheezed. (The SF is the Supreme Fascist, the Number-One Guy Up There, God, who was always tormenting Erdős by hiding his glasses, stealing his Hungarian passport, or, worse yet, keeping to Himself the elegant solutions to all sorts of intriguing mathematical problems.) "The SF created us to enjoy our suffering," Erdős said. "The sooner we die, the sooner we defy His plans."

Erdős still didn't see his friends, but his anger dissipated—his arm dropped to his side—as he heard the high-pitched squeal of a small boy, who was dining with his parents. "An epsilon!" Erdős said. (*Epsilon* was Erdős's word for a small child; in mathematics that Greek letter is used to represent small quantities.) Erdős moved slowly toward the child, navigating not so much by sight as by the sound of the boy's voice. "Hello," he said, as he reached into his ratty gray overcoat and extracted a bottle of Benzedrine. He dropped the bottle from shoulder height and with the same hand caught it a split second later. The epsilon was not at all amused, but perhaps to be polite, his parents made a big production of applauding. Erdős repeated the trick a few more times, and then he was rescued by one of his confederates, Ronald Graham, a mathematician at AT&T, who called him over to the table where he and Erdős's other friends were waiting.

The waitress arrived, and Erdős, after inquiring about each item on the long menu, ordered fried squid balls.

While the waitress took the rest of the orders, Erdős turned over his placemat and drew a tiny sketch vaguely resembling a rocket passing through a hula-hoop. His four dining companions leaned forward to get a better view of the world's most prolific mathematician plying his craft. "There are still many edges that will destroy chromatic number three," Erdős said. "This edge destroys bipartiteness." With that pronouncement Erdős closed his eyes and seemed to fall asleep.

Mathematicians, unlike other scientists, require no laboratory equipment—a practice that reportedly began with Archimedes, who, after emerging from his bath and rubbing himself with olive oil, discovered the principles of geometry by using his fingernails to trace figures on his oily skin. A Japanese restaurant, apparently, is as good a place as any to do mathematics. Mathematicians need only peace of mind and, occasionally, paper and pencil. "That's the beauty of it," Graham said. "You can lie back, close your eyes, and work. Who knows what problem Paul's thinking about now?"

"There was a time at Trinity College, in the 1930s I believe, when Erdős and my husband, Harold, sat thinking in a public place for more than an hour without uttering a single word," recalled Anne Davenport, the widow of one of Erdős's English collaborators. "Then Harold broke the long silence, by saying, 'It is not nought. It is one.' Then all was relief and joy. Everyone around them thought they were mad. Of course, they were."

■

Before Erdős died, on September 20, 1996, at the age of eighty-three, he had managed to think about more problems than any other mathematician in history. He wrote or co-authored 1,475 academic papers, many of them monumental, and all of them substantial. It wasn't just the quantity of work that was impressive but the quality: "There is an old saying," said Erdős. *"Non numerantur, sed ponderantur* (They are not counted but weighed). In the old [Hungarian] parliament of noblemen, they didn't count the votes: they weighed them. And this is true of papers. You know, Riemann had a very short list of papers, Gödel had a short list. Gauss was very prolific, as was Euler, of course." Even in his seventies there were years when Erdős published fifty papers, which is more than most good mathematicians write in a lifetime. He proved that mathematics isn't just a young man's game.

Erdős (pronounced "air-dish") structured his life to maximize the amount of time he had for mathematics. He had no wife or children, no job, no hobbies, not even a home, to tie him down. He lived out of a shabby suitcase and a drab orange plastic bag from Centrum Aruhaz ("Central Warehouse"), a large department store in Budapest. In a never-ending search for good mathematical problems and fresh mathematical talent, Erdős crisscrossed four continents at a frenzied pace, moving from one university or research center to the next. His modus operandi was to show up on the doorstep of a fellow mathematician, declare, "My brain is open," work with his host for a day or two, until he was bored or his host was run down, and then move on to another home.

Erdős's motto was not "Other cities, other maidens" but "Another roof, another proof." He did mathematics in more than twenty-five different countries, completing important proofs in remote places and sometimes publishing them in equally obscure journals. Hence the limerick, composed by one of his colleagues:

A conjecture both deep and profound
Is whether the circle is round.
In a paper of Erdős
Written in Kurdish
A counterexample is found.

When Erdős heard the limerick, he wanted to publish a paper in Kurdish but couldn't find a Kurdish math journal.

∎

Erdős first did mathematics at the age of three, but for the last twenty-five years of his life, since the death of his mother, he put in nineteen-hour days, keeping himself fortified with 10 to 20 milligrams of Benzedrine or Ritalin, strong espresso, and caffeine tablets. "A mathematician," Erdős was fond of saying, "is a machine for turning coffee into theorems." When friends urged him to slow down, he always had the same response: "There'll be plenty of time to rest in the grave."

Erdős would let nothing stand in the way of mathematical progress. When the name of a colleague in California came up at breakfast in New Jersey, Erdős remembered a mathematical result he wanted to share with him. He headed toward the phone and started to dial. His host interrupted him, pointing out that it was 5:00 A.M. on the West Coast. "Good," Erdős said, "that means he'll be home."

When challenged further in situations like this, Erdős was known to respond, "Louis the Fourteenth said, 'I am the state'; Trotsky could have said, 'I am society'; and I say, 'I am reality.'" No one who knew him would disagree. "Erdős had a childlike tendency to make his reality overtake yours," a friend said. "And he wasn't an easy houseguest. But we all wanted him around—for his mind. We all saved problems up for him."

To communicate with Erdős you had to learn his language. "When we met," said Martin Gardner, the mathematical essayist, "his first question was 'When did you arrive?' I looked at my watch, but Graham whispered to me that it was Erdős's way of asking, 'When were you born?' " Erdős often asked the same question another way: "When did the misfortune of birth overtake you?" His language had a special vocabulary—not just "the SF" and "epsilon" but also "bosses" (women), "slaves" (men), "captured" (married), "liberated" (divorced), "recaptured" (remarried), "noise" (music), "poison" (alcohol), "preaching" (giving a mathematics lecture), "Sam" (the United States), and "Joe" (the Soviet Union). When he said someone had "died," Erdős meant that the person had stopped doing mathematics. When he said someone had "left," the person had died.

·

At five foot six, 130 pounds, Erdős had the wizened, cadaverous look of a drug addict, but friends insist he was frail and gaunt long before he started taking amphetamines. His hair was white, and corkscrew-shaped whiskers shot out at odd angles from his face. He usually wore a gray pinstriped jacket, dark trousers, a red or mustard shirt or pajama top, and sandals or peculiar pockmarked Hungarian leather shoes, made especially for his flat feet and weak tendons. His whole wardrobe fit into his one small suitcase, with plenty of room left for his dinosaur of a radio. He had so few clothes that his hosts found themselves washing his socks and underwear several times a week. "He could buy more," one of his colleagues said, "or he could wash them himself. I mean, it takes zero IQ to learn how to operate a washing machine." But if it wasn't mathematics, Erdős wouldn't be bothered. "Some French socialist said that private property was theft," Erdős recalled. "I say that private property is a nuisance."

The only possessions that mattered to him were his mathematical notebooks. He filled ten of them by the time he died. He always carried one around with him, so that he could record his mathematical insights on a moment's notice. "Erdős came to my twins' bar mitzvah, notebook in hand," said Peter Winkler, a colleague of Graham's at AT&T. "He also brought gifts for my children—he loved kids—and behaved himself very well. But my mother-in-law tried to throw him out. She thought he was some guy who wandered in off the street, in a rumpled suit, carrying a pad under his arm. It is entirely possible that he proved a theorem or two during the ceremony."

All of his clothes, including his socks and custom-made underwear, were silk, because he had an undiagnosed skin condition that was aggravated by other kinds of fabric. He didn't like people to touch him. If you extended your hand, he wouldn't shake it. Instead, he'd limply flop his hand on top of yours. "He hated it if I kissed him," said Magda Fredro, a first cousin who was otherwise very close to him. "And he'd wash his hands fifty times a day. He got water everywhere. It was hell on the bathroom floor."

Although Erdős avoided physical intimacy, and was always celibate, he was friendly and compassionate. "He existed on a web of trust," said Aaron Meyerowitz, a mathematician at Florida Atlantic University. "When I was a graduate student and we had never met before, I gave him a ride. I didn't know the route and asked him if he wanted to navigate with a map. He didn't want to [and probably didn't know how to]. He just trusted that I, a total stranger, would get him there."

What little money Erdős received in stipends or lecture fees he gave away to relatives, colleagues, students, and strangers. He could not pass a homeless person without giving him money. "In the early 1960s, when I was a student at University College London," recalled D. G. Larman, "Erdős came to visit us for a year. After collecting his first month's salary he was accosted by a beggar on Euston station, asking for the price of a cup of tea. Erdős removed a small amount from the pay packet to cover his own frugal needs and gave the remainder to the beggar." In 1984 he won the prestigious Wolf Prize, the most lucrative award in mathematics. He contributed most of the $50,000 he received to a scholarship in Israel he established in the name of his parents. "I kept only seven hundred and twenty dollars," Erdős said, "and I remember someone commenting that for me even that was a lot of money to keep." Whenever Erdős learned of a good cause—a struggling classical music radio station, a fledgling Native American movement, a camp for wayward boys—he promptly made a small donation. "He's been gone a year," said Graham, "and I'm still getting mail from organizations he gave donations to. Today I got a postcard from an Israeli girls' home."

In the late 1980s Erdős heard of a promising high school student named Glen Whitney who wanted to study mathematics at Harvard but was a little short of the tuition. Erdős arranged to see him and, convinced of the young man's talent, lent him $1,000. He asked Whitney to pay him back only when it would not cause financial strain. A decade later Graham heard from Whitney, who at last had the money to repay Erdős. "Did Erdős expect me to pay interest?" Whitney wondered. "What should I do?" he asked Graham. Graham talked to Erdős. "Tell him," Erdős said, "to do with the thousand dollars what I did."

■

Erdős was a mathematical prodigy. At three he could multiply three-digit numbers in his head, and at four he discovered negative numbers. "I told my mother," he recalled, "that if you take 250 from 100, you get −150. My second great discovery was death. Children don't think they're ever going to die. I was like that too, until I was four. I was in a shop with my mother and suddenly I realized I was wrong. I started to cry. I knew I would die. From then on, I've always wanted to be younger. In 1970, I preached in Los Angeles on 'my first two and a half billion years in mathematics.' When I was a child, the Earth was said to be two billion years old. Now scientists say it's four and a half billion. So that makes me two and a half billion. The students at the lecture drew a timeline that showed me riding a dinosaur. I was asked, 'How were the dinosaurs?' Later, the right answer occurred to me: 'You know, I don't remember, because an old man only remembers the very early years, and the dinosaurs were born yesterday, only a hundred million years ago.'"

Erdős loved the dinosaur story and repeated it again and again in his mathematical talks. "He was the Bob Hope of mathematics, a kind of vaudeville performer who told the same jokes and the same stories a thousand times," said Melvyn Nathanson at a mathematical memorial service for Erdős in Budapest. "When he was scheduled to give yet another talk, no matter how tired he was, as soon as he was introduced to an audience, the adrenaline (or maybe amphetamine) would release into his system and he would bound onto the stage, full of energy, and do his routine for the thousand and first time."

In the early 1970s, Erdős started appending the initials P.G.O.M. to his name, which stood for Poor Great Old Man. When he turned sixty, he became P.G.O.M.L.D., the L.D. for Living Dead. At sixty-five he graduated to P.G.O.M.L.D.A.D., the A.D. for Archeological Discovery. At seventy he became P.G.O.M.L.D.A.D.L.D., the L.D. for Legally Dead. And at seventy-five he was P.G.O.M.L.D.A.D.L.D.C.D., the C.D. for Counts Dead. In 1987, when he was seventy-four, he explained: "The Hungarian Academy of Sciences has two hundred members. When you turn seventy-five, you can stay in the academy with full privileges, but you no longer count as a member. That's why the C.D. Of course, maybe I won't have to face that emergency. They are planning an international conference for my seventy-fifth birthday. It may have to be for my memory. I'm miserably old. I'm really not well. I don't understand what's happening to my body—maybe the final solution."

Erdős outlived most of his friends and, to his dismay, watched some of them lose their minds. His college thesis adviser, Leopold Féjer, one of the strongest mathematicians in Hungary, was burned out by the age of thirty. "He still did very good things, but he felt that he didn't have any significant new ideas," said Erdős. "When he was sixty, he had a prostate operation and after that he didn't do very much. Then he was on an even keel for fifteen or sixteen years, and then he became senile. There was some disturbance of the circulation. It was very sad because he knew he was senile and he said things like, 'Since I became a complete idiot. . . .' He was very well kept in the hospital but died of a stroke in 1959."

When Paul Turán, his closest friend, with whom he

had written thirty papers, died in 1976, Erdős had an image of the SF assessing the work he had done with his collaborators. On one side of a balance the SF would place the papers Erdős had co-authored with the dead; on the other side the papers written with the living. "When the dead side tips the balance," Erdős said, "I must die too." He paused for a moment and then added, "It's just a joke of mine."

Perhaps. But for decades Erdős vigorously sought out new, young collaborators and ended many working sessions with the remark, "We'll continue tomorrow if I live." With 485 co-authors, Erdős collaborated with more people than any other mathematician in history. Those lucky 485 are said to have an Erdős number of 1, a coveted code phrase in the mathematics world for having written a paper with the master himself. If your Erdős number is 2, it means you have published with someone who has published with Erdős. If your Erdős number is 3, you have published with someone who has published with someone who has published with Erdős. Einstein had an Erdős number of 2, and the highest known Erdős number of a working mathematician is 7. The great unwashed who have never written a mathematical paper have an Erdős number of ∞.

"I was told several years ago that my Erdős number was 7," Caspar Goffman at Purdue wrote in 1969. But "it has recently been lowered to 3. Last year I saw Erdős in London. . . . When I told him the good news that my Erdős number had just been lowered, he expressed regret that he had to leave London that same day. Otherwise an ultimate lowering might have been accomplished."

With Erdős's death, the No. 1 Club's membership will hardly grow, except for the admission of a few stragglers who had joint papers with him in the works that should be published soon. "When these papers come out," said Graham, "we'll scrutinize them carefully to make sure no one is pretending to have worked with Erdős." And those who could have worked with him but didn't are having regrets. "One evening in the seventies," recalled MIT mathematician Gian-Carlo Rota, "I mentioned to Paul a problem in numerical computation I was working on. Instantly, he gave me a hint that eventually led to the complete solution. We thanked him for his help in the introduction to our paper, but I will always regret not having included his name as a co-author. My Erdős number will now permanently remain equal to two."

The mathematical literature is peppered with tongue-in-cheek papers probing the properties of Erdős numbers, and Jerrold Grossman at Oakland University in Roches-ter, Michigan, runs an Internet site, called the Erdős Number Project, which tracks the coveted numbers. One of Erdős's specialties was graph theory. By a *graph*, math-ematicians don't mean the kind of chart Ross Perot waved at the TV cameras. They mean any group of points ("vertices" is the lingo) connected by lines ("edges"). So a triangle, for example, is a graph with three vertices and three edges. Now take Erdős's 485 col-laborators and represent them by 485 points on a sheet of paper. Draw an edge between any two points whenever the corresponding mathematicians published together. The resulting graph, which at last count had 1,381 edges, is the Collaboration Graph.

Some of Erdős's colleagues have published papers about the properties of the Collaboration Graph, treating it as if it were a real mathematical object. One of these papers made the observation that the graph would have a certain very interesting property if two particular points had an edge between them. To make the Collaboration Graph have that property, the two disconnected mathematicians immediately got together, proved something trivial, and wrote up a joint paper.

"I wrote a paper once about the Collaboration Graph," said Graham, "that filled, I claimed, a much-needed gap in the mathematical literature. Well, if the gap was much needed, I shouldn't have written the paper!" There is a tradition of writing these papers under pseudonyms. "I've used the name Tom Odda," said Graham. Tom Odda? "Look it up in *Maledicta, the Journal of Verbal Aggression,*" said Graham. "You'll find it under Mandarin terms of abuse. Tom Odda means *your mother's*____, where the blank is too unmentionable even for *Maledicta* to fill in."

■

Though he was confident of his skill in mathematics, outside that arcane world Erdős was very nearly helpless. After his mother's death, the responsibility of looking after him fell chiefly to Ronald Graham, who spent almost as much time in the 1980s handling Erdős's affairs as he did overseeing the seventy mathematicians, statisticians, and computer scientists at AT&T Bell Labs. Graham was the one who called Washington when the SF stole Erdős's visa; and during Erdős's last few years, he said, "the SF struck with increasing frequency." Graham also managed Erdős's money, and was forced to become an expert on currency exchange rates because honoraria from Erdős's lectures dribbled in from four continents. "I signed his name on checks and deposited them," Graham said. "I did this so long I doubt the bank would have cashed a check if he had endorsed it himself."

On the wall of Graham's old office, in Murray Hill,

New Jersey, was a sign: ANYONE WHO CANNOT COPE WITH MATHEMATICS IS NOT FULLY HUMAN. AT BEST HE IS A TOLERABLE SUBHUMAN WHO HAS LEARNED TO WEAR SHOES, BATHE, AND NOT MAKE MESSES IN THE HOUSE. Near the sign was the "Erdős Room," a closet full of filing cabinets containing copies of more than a thousand of Erdős's articles. "Since he had no home," Graham said, "he depended on me to keep his papers, his mother having done it earlier. He was always asking me to send some of them to one person or another." Graham also handled all of Erdős's incoming correspondence, which was no small task, because many of Erdős's mathematical collaborations took place by mail. He sent out 1,500 letters a year, few of which dwelt on subjects other than mathematics. "I am in Australia," a typical letter began. "Tomorrow I leave for Hungary. Let $k$ be the largest integer. . . ."

Graham had less success influencing Erdős's health. "He badly needed a cataract operation," Graham said. "I kept trying to persuade him to schedule it. But for years he refused, because he'd be laid up for a week, and he didn't want to miss even seven days of working with mathematicians. He was afraid of being old and helpless and senile." Like all of Erdős's friends, Graham was concerned about his drug-taking. In 1979, Graham bet Erdős $500 that he couldn't stop taking amphetamines for a month. Erdős accepted the challenge, and went cold turkey for thirty days. After Graham paid up—and wrote the $500 off as a business expense—Erdős said, "You've showed me I'm not an addict. But I didn't get any work done. I'd get up in the morning and stare at a blank piece of paper. I'd have no ideas, just like an ordinary person. You've set mathematics back a month." He promptly resumed taking pills, and mathematics was the better for it.

In 1987 Graham built an addition onto his house in

Watchung, New Jersey, so that Erdős would have his own bedroom, bathroom, and library for the month or so he was there each year. Erdős liked staying with Graham because the household contained a second strong mathematician, Graham's wife, Fan Chung, a Taiwanese émigré who today is a professor at the University of Pennsylvania. When Graham wouldn't play with him, Chung would, and the two co-authored thirteen papers, the first in 1979.

Back in the early 1950s, Erdős started spurring on his collaborators by putting out contracts on problems he wasn't able to solve. By 1987, the outstanding rewards totaled about $15,000 and ranged from $10 to $3,000, reflecting his judgment of the problems' difficulty. "I've had to pay out three or four thousand dollars," Erdős said then. "Someone once asked me what would happen if all the problems were solved at once. Could I pay? Of course I couldn't. But what would happen to the strongest bank if all the creditors asked for their money back? The bank would surely go broke. And a run on the bank is much more likely than solutions to all my problems." Now that he's gone, Graham and Chung have decided to pay the cash prizes themselves for Erdős's problems in graph theory, about which they have published a book. More than one hundred graph theory problems have a contract on them, for a total of more than $10,000. Andrew Beal, a Dallas banker and amateur mathematician, has offered to help bankroll Erdős's problems in other fields.

Graham and Erdős would seem an unlikely pair. Although Graham is one of the world's leading mathematicians, he did not, like Erdős, forsake body for mind. Indeed, he continues to push both to the limit. At six foot two, with blond hair, blue eyes, and chiseled features, Graham looks at least a decade younger than his sixty-two years. He can juggle six balls and is a past president of the International Jugglers Association. He is an accomplished trampolinist, who put himself through college as a circus acrobat. ("Trampolining is just like juggling," said Graham, exhibiting a mathematician's tendency to generalize, "only there's just one object to juggle—yourself.") He has bowled two 300 games, is vicious with a boomerang, and more than holds his own at tennis and Ping-Pong.

While Erdős could sit for hours, Graham is always moving. In the middle of solving a mathematical problem he'll spring into a handstand, grab stray objects and juggle them, or jump up and down on the super-springy pogo stick he keeps in his office. "You can do mathematics anywhere," Graham said. "I once had a flash of insight into a stubborn problem in the middle of a back somersault with a triple twist on my trampoline."

"If you add up Ron's mathematical theorems and his double somersaults," one of his colleagues said, "he'd surely have a record." Graham, in fact, does hold a world record— one no less peculiar. He was cited in *The Guinness Book of World Records* for coming up with the largest number ever used in a mathematical proof. The number is incomprehensibly large. Mathematicians often try to suggest the magnitude of a large number by likening it to the number of atoms in the universe or the number of grains of sand in the Sahara. Graham's number has no such physical analogue. It can't even be expressed in familiar mathematical notation, as, say, the number 1 followed by a zillion zeroes. To cite it, a special notation had to be invoked in which exponents are heaped on exponents to form a staggering leaning tower of digits.

Besides staying on the cutting edge of mathematics and acrobatics, Graham found time to learn Chinese and take up the piano. Neither his wife nor his coworkers understand how he does it. "It's easy," Graham said. "There are a hundred and sixty-eight hours in *every* week."

Erdős and Graham met in 1963 in Boulder, Colorado, at a conference on number theory, and immediately began collaborating, writing twenty-seven papers and one book together. That meeting was also the first of many spirited athletic encounters the two men had. "I remember thinking when we met that he was kind of an old guy," Graham said, "and I was amazed that he beat me at Ping-Pong. That defeat got me to take up the game seriously." Graham bought a machine that served Ping-Pong balls at very high speeds and went on to become Ping-Pong champion of Bell Labs. Even when Erdős was in his eighties, they still played occasionally. "Paul loved challenges," said Graham. "I'd give him nineteen points and play sitting down. But his eyesight was so bad that I could just lob the ball high into the air and he'd lose track of it."

In later years Erdős came up with novel athletic contests at which he'd seem to have more of a chance, though he invariably lost. "Paul liked to imagine situations," Graham said. "For example, he wondered whether I could climb stairs twice as fast as he could. We decided to see. I ran a stopwatch as we both raced up twenty flights in an Atlanta hotel. When he got to the top, huffing, I punched the stopwatch but accidentally erased the times. I told him we'd have to do it again. 'We're *not* doing it again,' he growled, and stormed off.

"Another time, in Newark Airport, Erdős asked me how hard it was to go up a down escalator. I told him it could be done, and I demonstrated. 'That was harder than I thought,' I said. 'That looks easy,' he said. 'I'm sure you couldn't do it,' I said. 'That's ridiculous,' he said. 'Of course I can.' Erdős took about four steps up the escalator and then fell over on his stomach and slid down to the bottom. People were staring at him. He was wearing this ratty coat and looked like he was a wino from the Bowery. He was indignant afterward. 'I got dizzy,' he said."

Erdős and Graham were like an old married couple, happy as clams but bickering incessantly, following scripts they knew by heart though they were baffling to outsiders. Many of these scripts centered on food. When Erdős was feeling well, he got up about 5:00 A.M. and started banging around. He'd like Graham to make him breakfast, but Graham thought he should make his own. Erdős loved grapefruit, and Graham stocked the refrigerator when he knew Erdős was coming. On a visit in the spring of 1987, Erdős, as always, peeked into the refrigerator and saw the fruit. In fact, each knew that the other knew that the fruit was there.

"Do you have any grapefruit?" Erdős asked.

"I don't know," Graham replied. "Did you look?"

"I don't know where to look."

"How about the refrigerator?"

"Where in the refrigerator?"

"Well, just look."

Erdős found a grapefruit. He looked at it and looked at it and got a butter knife. "It can't be by chance," Graham explained, "that he so often used the dull side of the knife, trying to force his way through. It'll be squirting like mad, all over himself and the kitchen. I'd say, 'Paul, don't you think you should use a sharper knife?' He'd say, 'It doesn't matter,' as the juice shoots across the room. At that point I give up and cut it for him."

Graham was not the only one who had to put up with
Erdős's kitchen antics. "Once I spent a few days with Paul,"
said János Pach, a fellow Hungarian émigré. "When I entered the kitchen in the evening, I was met with a horrible
sight. The floor was covered by pools of blood-like red liq-
uid. The trail led to the refrigerator. I opened the door,
and to my great surprise saw a carton of tomato juice on
its side with a gaping hole. Paul must have felt thirsty and,
after some reflection, decided to get the juice out of the
carton by stabbing it with a big knife."

In mathematics, Erdős's style was one of intense curiosity, a style he brought to everything else he confronted.
Part of his mathematical success stemmed from his will-
ingness to ask fundamental questions, to ponder critically
things that others had taken for granted. He also asked
basic questions outside mathematics, but he never remembered the answers, and asked the same questions again and
again. He'd point to a bowl of rice and ask what it was
and how it was cooked. Graham would pretend he didn't
know; others at the table would patiently tell Erdős about
rice. But a meal or two later Erdős would be served rice
again, act as though he'd never seen it, and ask the same
questions.

Erdős's curiosity about food, like his approach to so
many things, was merely theoretical. He never actually
tried to cook rice. In fact, he never cooked anything at all,
or even boiled water for tea. "I can make excellent cold
cereal," he said, "and I could probably boil an egg, but I've
never tried." He was twenty-one when he buttered his first
piece of bread, his mother or a domestic servant having
always done it for him. "I remember clearly," he said. "I
had just gone to England to study. It was teatime, and

bread was served. I was too embarrassed to admit that I had never buttered it. I tried. It wasn't so hard." Only ten years before, at the age of eleven, he had tied his shoes for the first time.

His curiosity about driving was legendary in the mathematics community, although you never found him behind the wheel. He didn't have a license and depended on a network of friends, known as "Uncle Paul sitters," to chauffeur him around. But he was constantly asking what street he was on and questioning whether it was the right one. "He was not a nervous wreck," Graham said. "He just wanted to know. Once he was driving with Paul Turán's widow, Vera Sós. She had just learned to drive, and Paul was doing his usual thing, 'What about this road?' 'What about that road?' 'Shouldn't we be over there?' Vera was distracted and she plowed into the side of a car that must have been going forty or fifty miles an hour. She totaled it, and vowed that she would never drive with Erdős again."

But outside mathematics, Erdős's inquisitiveness was limited to necessities like eating and driving; he had no time for frivolities like sex, art, fiction, or movies. Erdős last read a novel in the 1940s, and it was in the 1950s that he apparently saw his last movie, *Cold Days*, the story of an atrocity in Novi Sad, Yugoslavia, in which Hungarians brutally drowned several thousand Jews and Russians. Once in a while the mathematicians he stayed with forced him to join their families on nonmathematical outings, but he accompanied them only in body. "I took him to the Johnson Space Center to see rockets," one of his colleagues recalled, "but he didn't even look up." Another mathematician took him to see a mime troupe, but he fell asleep before the performance started. Melvyn Nathanson, whose wife was a curator at the Museum of Modern Art in New York, dragged Erdős there. "We showed him Matisse," said Nathanson, "but he would have nothing to do with it. After a few minutes we ended up sitting in the Sculpture Garden doing mathematics."

# 1

## STRAIGHT FROM THE BOOK

I have wished to understand the hearts of men. I have wished to know why the stars shine. And I have tried to apprehend the Pythagorean power by which number holds sway above the flux.

—Bertrand Russell

The mathematician's patterns, like the painter's or the poet's, must be beautiful; the ideas, like the colours or the words, must fit together in a harmonious way. Beauty is the first test: there is no permanent place in the world for ugly mathematics. . . . It may be very hard to define mathematical beauty, but that is just as true of beauty of any kind—we may not know quite what we mean by a beautiful poem, but that does not prevent us from recognizing one when we read it.

—G. H. Hardy

Erdős was a mathematical monk. He renounced physical pleasure and material possessions for an ascetic, contemplative life, a life devoted to a single narrow mission: uncovering mathematical truth. What was this mathematics that could possibly be so diverting and consuming?

"There's an old debate," Erdős said, "about whether you create mathematics or just discover it. In other words, are the truths already there, even if we don't yet know them? If you believe in God, the answer is obvious. Mathematical truths are there in the SF's mind, and you just rediscover them. Remember the limericks:

> There was a young man who said, 'God,
> It has always struck me as odd
> That the sycamore tree
> Simply ceases to be
> When there's no one about in the quad.'

> 'Dear Sir, Your astonishment's odd;
> I am always about in the quad:
> And that's why the tree
> Will continue to be,
> Since observed by,
> Yours faithfully, God.'

"I'm not qualified to say whether or not God exists," Erdős said. "I kind of doubt He does. Nevertheless, I'm always saying that the SF has this transfinite Book—transfinite being a concept in mathematics that is larger than infinite—that contains the best proofs of all mathematical theorems, proofs that are elegant and perfect." The strongest compliment Erdős gave to a colleague's work was to say, "It's straight from the Book."

"I was once introducing Erdős at a lecture," said Joel Spencer, a mathematician at New York University's Courant Institute who began working with Erdős in 1970. "And I started to talk about his idea of God and the Book. He interrupted me and said, 'You don't have to believe in God, but you should believe in the Book.' Erdős has made me and other mathematicians recognize the importance of what we do. Mathematics is there. It's beautiful. It's this jewel we uncover."

That mathematics could be a jewel might come as a surprise to those who struggled with multiplication tables as kids and now need help completing income tax forms. Mathematics is a misunderstood and even maligned discipline. It's not the brute computations they drilled into us in grade school. It's not the science of reckoning. Mathematicians do not spend their time thinking up cleverer ways of multiplying, faster methods of adding, better schemes for extracting cube roots.

■

When I was in grade school, I thought mathematics was about brute computation—although I confess that I liked computation because I was good at it. My math teacher would ask two of us to stand. Then she would pose an arithmetic problem that we would race to solve in our heads. The first to blurt out the correct answer would remain standing and the loser would sit down. A challenger would then rise, and the teacher would pose another problem. This went on for the entire class. More often than not I remained standing the whole time. (By junior high school, however, my interest in computation had taken a backseat to another subject with an equally rigid set of rules: chess. I played it day and night, with others and by myself. For a few years I was as obsessed with chess as Erdős was with math—I played out chess games move for move in my sleep—but unfortunately I didn't have anywhere near the results Erdős had.)

My first job after college was at *Scientific American*,

and for a short while I found myself editing Martin Gardner's well-known "Mathematical Games" column. Gardner was somewhat of a recluse who seldom came into the office, but from his writings I learned that mathematics was much more than arithmetic. I remember being captivated by the head-spinning idea that there were infinities larger than infinity. Through *Scientific American* I met Ronald Graham in 1980 and through Graham I met Erdős several years later. Their devotion to mathematics was infectious, and how they did mathematics defied the stereotype of the asocial genius. Neither one holed himself up in a musty study, proving and conjecturing with no other souls around. For Erdős and Graham, mathematics was a group activity. I watched them argue passionately about mathematical ideas, as I had seen them fight about grapefruit.

I wanted to understand their world. I sought out people with an Erdős number 1. I talked to the spouses. I slept in Erdős's bedroom in Graham's house (I'm not sure what I expected but the experience did nothing to improve my mathematical ability). I immersed myself in the history of mathematical ideas. I studied Pythagoras, Newton, Fermat, Gauss, Hilbert, Einstein, and Gödel. I read mathematical memoirs, pored over Erdős's correspondence, peeked in his lone suitcase, and conversed with him at various times over the period of a decade. I grew fond of him, laughed at his silly quips, and came to appreciate why he saw mathematics as the search for lasting beauty and ultimate truth. It was a search, I learned, that he never lost sight of even when his life was torn asunder by major political dramas of the twentieth century—the Communist revolutions in Hungary, the rise of Fascism and anti-Semitism in Europe, World War II, the Cold War, McCarthyism. Mathematics was his anchor in a world that he regarded as cruel and

heartless, although he believed in the goodness and innocence of ordinary individuals. "The game of life," Erdős
often said, "is to keep the SF's score low. If you do something bad in life, the SF gets two points. If you don't do
something good that you should have done, the SF gets one
point. You *never* score, so the SF always wins." But the aim
of life, he emphasized, is to prove and conjecture. "Mathematics is the surest way to immortality. If you make a big
discovery in mathematics, you will be remembered after
everyone else will be forgotten."

■

Even people drawn to mathematics have had misconcep-
tions about its scope. "I always wanted to be a mathematician," Spencer said, "even before I knew what
mathematicians did. My father was a CPA, and I loved
numbers. I thought mathematics was about adding up
longer and longer lists. I found out what it really was in
high school. I'd undoubtedly be a lot richer now if I were
making my living adding up long lists of numbers."

János Pach, now a geometer at New York University,
remembers as a child watching his aunt Vera Sós and Paul
Turán work together with Erdős in the late 1950s at a
guesthouse of the Hungarian Academy of Sciences: "When
the grown-ups went for a walk, or played Ping-Pong, or
were having coffee, and the scene became quiet, I sneaked
over to the table to catch a glance at 'higher mathematics,'
i.e., at the notes scattered on the table. I was astonished
when I first saw the end-product of their work: strange
letters, numbers, signs, arrows, scribble-scrabble. . . . I had
no doubt that the Laws of the Universe were written in
this mysterious language. Otherwise, how could mathematical problems spark such enthusiasm in these brilliant

and famous people!" Pach was also drawn to mathematics because of the air of success of these grown-ups. They "traveled all over the world. They lectured at every important university from Beijing to Calgary. They had a worldly air about them. They owned fine tweed jackets, listened to pocket radios, and wore shoes that required no laces! Such things do not escape a teenager's attention."

But such things did not impress Erdős's cousin Magda Fredro, who hadn't the slightest idea what he did, even though she knew him for most of his life and accompanied him on mathematical sojourns from Florida to Israel. "Tell me, what is this about?" she asked me, flipping through her copy of Erdős's book *The Art of Counting*. "It looks like Chinese. Also, tell me, how famous and brilliant is he? I know so little about him. He once looked up six phone numbers. Then we talked for half an hour before he phoned them all, from memory. More than all his scientific work, that impressed me."

Mathematics, in its abstractness and observation of formal rules, has been likened to chess. Both activities demand of their practitioners deep concentration, the tuning out of one's surroundings to focus on the formal structure at hand, and a kind of "can-do" mindset. "It is most important in creative science not to give up," said the Polish-born mathematician Stanislaw Ulam. "If you are an optimist you will be willing to 'try' more than if you are a pessimist. It is the same in games like chess. A really good chess player tends to believe (sometimes mistakenly) that he holds a better position than his opponent. This, of course, helps to keep the game moving and does not increase the fatigue that self-doubt engenders."

But analogies to chess should only be pushed so far. "A chess problem is genuine mathematics, but it is in some way 'trivial' mathematics," wrote the number theorist G. H.

Hardy in his classic book *A Mathematician's Apology*. "However ingenious and intricate, however original and surprising the moves, there is something essential lacking. Chess problems are unimportant. The best mathematics is *serious* as well as beautiful." And it should be emphasized that "no chess problem has ever affected the general development of scientific thought; Pythagoras, Newton, Einstein have in their times changed its whole direction."

For Erdős, Graham, and their colleagues, mathematics is order and beauty at its purest, order that transcends the physical world. When Euclid, the Greek geometer of the third century B.C., spoke of points and lines, he was speaking of idealized entities, points that have no dimension and lines that have no width. All points and lines that exist in the real world—in, say, physics or engineering—do have dimension and thus are only imperfect imitations of the pure constructs that geometers ponder. Only in this idealized world do the angles of every triangle always sum to precisely 180 degrees.

Numbers, too, can have a transcendent quality. Take the prime numbers, integers like 2, 3, 5, 7, 11, 13, and 17, which are evenly divisible only by themselves and the number 1. We happen to have ten fingers, and our number system is conveniently based on ten digits. But the same primes, with all the same properties, exist in any number system. If we had twenty-six fingers and constructed our number system accordingly, there would still be primes. The universality of primes is the key to Carl Sagan's novel *Contact*, in which extraterrestrials, with God only knows how many fingers, signal Earthlings by emitting radio signals that have a prime number of pulses. But little green men need not be invoked in order to conceive of a culture that doesn't use base 10. We have had plenty here on Earth. Computers use a binary system, and the Babylonians had a base-60 system, vestiges of which are evident in the way we measure time (sixty seconds in a minute, sixty minutes in an hour). Cumbersome as this sexagesimal system was, it, too, contains the same primes. So does the octary system that Reverend Hugh Jones, a mathematician at the College of William and Mary, championed in the eighteenth century as more natural for women than base-10 because of women's experience in the kitchen working with multiples of 8 (32 ounces in a quart, 16 ounces in a pound).

Numbers, Hardy believed, constituted the true fabric of the universe. In an address to a group of physicists in 1922, he took the provocative position that it is the mathematician who is in "much more direct contact with reality. This may seem a paradox, since it is the physicist who deals with the subject-matter usually described as 'real.'" But "a chair or a star is not in the least like what it seems to be; the more we think of it, the fuzzier its outlines become in the haze of sensation which surrounds it; but '2' or '317' has nothing to do with sensation, and its properties stand out the more clearly the more closely we scrutinize it . . . 317 is a prime, not because we think so, or because our minds are shaped in one way rather than another, but *because it is so*, because mathematical reality is built that way."

Prime numbers are like atoms. They are the building blocks of all integers. Every integer is either itself a prime or the product of primes. For example, 11 is a prime; 12 is the product of the primes 2, 2, and 3; 13 is a prime; 14 is the product of the primes 2 and 7; 15 is the product of the primes 3 and 5; and so on. Some 2,300 years ago, in proposition 20 of Book IX of his *Elements*, Euclid gave a proof, "straight from the Book," that the supply of primes is inexhaustible.

(Assume, said Euclid, that there is a finite number of

primes. Then one of them, call it $P$, will be the largest. Now consider the number $Q$, larger than $P$, that is equal to the product of the consecutive whole numbers from 2 to $P$ plus the number 1. In other words, $Q = (2 \times 3 \times 4 \ldots \times P) + 1$. From the form of the number $Q$, it is obvious that no integer from 2 to $P$ divides evenly into it; each division would leave a remainder of 1. If $Q$ is not prime, it must be evenly divisible by some prime larger than $P$. On the other hand, if $Q$ is prime, $Q$ itself is a prime larger than $P$. Either possibility implies the existence of a prime larger than the assumed largest prime $P$. This means, of course, that the concept of "the largest prime" is a fiction. But if there's no such beast, the number of primes must be infinite. "Euclid alone," wrote Edna St. Vincent Millay, "has looked on Beauty bare.")

As of this writing, the largest known prime is a 909,526-digit number formed by raising 2 to the 3,021,377th power and subtracting 1. The prime was found on January 27, 1998, by the GIMPS project (Great Internet Mersenne Prime Search), in which 4,000 "primees" (prime-number groupies) communicated over the Internet and pooled their computers for the hunt. Each of the 4,000 computers was assigned an interval of numbers to check. Roland Clarksen, a 19-year-old sophomore at California State University Dominguez Hills, was the lucky primee whose 200 Mhz Pentium-based home PC, after 46 days of running part-time, examining the numbers in his assigned interval, proved the primality of $2^{3,021,377} - 1$.

The hunt for large primes has come a long way since the seventeenth century, when Marin Mersenne, a Parisian monk, took time out from his monastic duties to search for primes. A number like $2^{3,021,377} - 1$ that is of the form $2^n - 1$ is said to be a *Mersenne number*. For a Mersenne number to be prime, $n$ itself must be prime. Thus, since

$2^{3,021,377} - 1$ is prime, 3,021,377 must also be prime. But $n$ being prime does not guarantee that the corresponding Mersenne number is prime. When $n$ takes on the first four prime numbers, Mersenne primes are indeed generated:

$$\text{For } n = 2, 2^2 - 1 = 3$$
$$\text{For } n = 3, 2^3 - 1 = 7$$
$$\text{For } n = 5, 2^5 - 1 = 31$$
$$\text{For } n = 7, 2^7 - 1 = 127$$

But when $n$ is the fifth prime number, 11, the corresponding Mersenne number proves to be composite ($2^{11} - 1$ is 2,047, whose prime factors are 23 and 89). In 1644, Mersenne himself claimed that when $n$ took on the values of the sixth, seventh, and eighth prime numbers, namely, 13, 17, and 19, the corresponding Mersenne numbers, $2^{13} - 1$ (or 8,191), $2^{17} - 1$ (or 131,071) and $2^{19} - 1$ (or 524,287) were primes. He was right.

The monk also made the bold claim that $2^{67} - 1$ was prime. The claim was not disputed for more than 250 years. Then, in 1903, Frank Nelson Cole of Columbia University delivered a talk with the unassuming title "On the Factorization of Large Numbers" at a meeting of the American Mathematical Society. "Cole—who was always a man of very few words—walked to the board and, saying nothing, proceeded to chalk up the arithmetic for raising 2 to the sixty-seventh power," recalled Eric Temple Bell, who was in the audience. "Then he carefully subtracted 1 [getting the 21-digit monstrosity 147,573,952,589,676,412,927]. Without a word he moved over to a clear space on the board and multiplied out, by longhand,

$$193,707,721 \times 761,838,257,287$$

"The two calculations agreed. Mersenne's conjecture—if such it was—vanished into the limbo of mathematical mythology. For the first . . . time on record, an audience of the American Mathematical Society vigorously applauded the author of a paper delivered before it. Cole took his seat without having uttered a word. Nobody asked him a question."

Prime numbers are elusive because no formula like Mersenne's $2^n - 1$ generates only primes. The technique the GIMPS project uses isn't much more sophisticated than a 2,000-year-old method called "the sieve of Eratosthenes," invented by Eratosthenes of Alexandria, whose nickname was "Beta" because he was said to be at least second best in everything, from geometry to drama. The idea of the sieve is simple. List consecutive positive integers starting with 2. Then cross off all multiples of the first prime, 2. What's left is the next prime, 3. Now cross off all multiples of 3. What's left is the next prime, 5. Now cross off all multiples of 5. Each successive "sieving" catches another prime.

For generations of mathematicians, prime numbers have always had an almost mystical appeal. "I even know of a mathematician who slept with his wife only on prime-numbered days," Graham said. "It was pretty good early in the month—two, three, five, seven—but got tough toward the end, when the primes are thinner, nineteen, twenty-three, then a big gap till twenty-nine. But this guy was seriously nuts. He's now serving twenty years in the Oregon State Penitentiary for kidnapping and attempted murder."

Prime numbers are appealing because, in spite of their apparent simplicity, their properties are extremely elusive. All sorts of basic questions about them remain unanswered, even though they have been scrutinized by generations of the sharpest mathematical minds. In 1742, for example, Christian Goldbach conjectured that every even number greater than 2 is the sum of two primes:

$$4 = 2 + 2$$
$$6 = 3 + 3$$
$$8 = 5 + 3$$
$$10 = 5 + 5$$
$$12 = 7 + 5$$
$$14 = 7 + 7$$

and so on.

"Descartes actually discovered this before Goldbach," said Erdős, "but it is better that the conjecture was named for Goldbach because, mathematically speaking, Descartes was infinitely rich and Goldbach was very poor." With the aid of computers, twentieth-century mathematicians have shown that all even numbers up to 100 million can be expressed as the sum of two primes, but they have not been able to prove that Goldbach's simple conjecture is universally true. Similarly, computer searches have revealed numerous *twin primes,* pairs of consecutive odd numbers both of which are prime: 3 and 5, 5 and 7, 11 and 13, 71 and 73, 1,000,000,000,061 and 1,000,000,000,063. Number theorists believe that the supply of twin primes is inexhaustible, like the supply of primes themselves, but no one has been able to prove this. On an even deeper level, no one has found an easy way of telling in advance how far one prime number will be from the next one.

That the gaps between primes have no known pattern served as a welcome distraction for the journalist Roger Cooper when he was kept in solitary confinement in Iran for several years in the 1980s: "Between interrogations, always blindfolded and accompanied by slaps and punches when I refused to confess to being a British spy, I tried to find ways of amusing myself without books. I made a backgammon set with dice of bread, and evolved a maths system based on Roman numerals but with an apple pip for zero. Orange pips were units, plum stones were fives, and tens and hundreds were positional. This enabled me to calculate all the prime numbers up to 5,000, which I recorded in dead space where the door opened and could speculate on the anomalies in their appearance."

∎

The prime numbers were Erdős's intimate friends. He understood them better than anyone else. "When I was ten," he said, "my father told me about Euclid's proof, and I was hooked." Eight years later, as a college freshman, he caused a stir in Hungarian mathematics circles with a simple proof that a prime can always be found between any integer (greater than 1) and its double. This result had already been proved in about 1850 by one of the fathers of Russian mathematics, Pafnuty Lvovitch Chebyshev. But Chebyshev's proof was too heavy-handed to be in the Book. He had used a steam shovel to transplant a rosebush, whereas Erdős managed with a spoon. News of Erdős's youthful triumph was spread through the English-speaking world by the ditty:

Chebyshev said it, and I say it again
There is always a prime between $n$ and $2n$

In 1939, Erdős attended a lecture at Princeton by Mark Kac, a Polish émigré mathematical physicist who would contribute to the American development of radar during World War II. "He half-dozed through most of my lecture," Kac recalled. "The subject matter was too far removed from his interests. Toward the end I described briefly my difficulties with the number of prime divisors. At the mention of number theory Erdős perked up and asked me to explain once again what the difficulty was. Within the next few minutes, even before the lecture was over, he interrupted to announce that he had the solution!"

Erdős's insight into primes was so keen that, on hearing a new problem about them, he often quickly upstaged those who had spent far more time thinking about it. In his autobiography *Eye of the Hurricane*, Richard Bellman tells the story of how he and Hal Shapiro addressed the problem of determining the probability that two numbers are "relatively prime" (meaning that they have no common divisors larger than 1). They wrote up their solution and submitted the manuscript to the *Transactions of the American Mathematical Society*. "Unfortunately, the editor, A. A. Albert, had gone off to Brazil for six months and left the journal in the hands of Kaplansky. Our paper was refereed by Erdős who accepted it. Kaplansky, however, asked Erdős whether he couldn't get a better proof than ours. Erdős supplied one, and Kaplansky transmitted it to us. We wrote back, pointing out if Erdős's proof was written up in the same detail as ours it would be just as long. Kaplansky persisted, and Erdős supplied a very elegant shorter proof. Kaplansky suggested then a joint paper. I refused, saying that we would add Erdős's paper as an appendix if he desired."

The stalemate remained for a month, until, Bellman recalled, "Shapiro came to me very disturbed. He had gone to various parties and it seems that Erdős had gone around the country saying that we were stealing our result from him." Shapiro and Bellman then wrote to Kaplansky, retracting the paper. Several weeks later they got a letter from Kaplansky, saying that the type had been destroyed and billing them $300 for the typesetting. Shapiro and Bellman refused to pay the bill. By that time Albert had returned and fired off "a vitriolic letter" threatening Bellman and Shapiro with expulsion from the American Mathematical Society if they did not pay up. "I have often thought of the expulsion ceremony," Bellman wrote. "I think our pencils would be broken in front of us."

At this point, Bellman appealed to Lefschetz, offering an eloquent argument about how the referee's job is to judge the truth and significance of a result, not to improve it. Lefschetz ignored the subtleties of the argument and "wrote a letter to Albert pointing out various indiscretions which Albert had committed in his youth." Albert then graciously backed down. Neither Bellman and Shapiro's paper nor Erdős's proof was ever published, which, Bellman noted, "is a great shame since Erdős really had an elegant method."

In 1949, Erdős had his greatest victory over the prime numbers, although the victory was one he didn't like to talk about because it too was marred by controversy. Although mathematicians have no effective way of telling exactly where prime numbers lie, they have known since the late eighteenth century of a formula that describes the statistical distribution of primes, how on average the primes thin out the further out you go. In search of such a formula, Carl Friedrich Gauss, arguably the greatest mathematician of all time, began studying tables of prime numbers in 1792, when he was fifteen. By 1796, Johann Heinrich Lambert and Georg von Vega had published a list of all the primes up to 400,031, and other investigators soon extended the list into the millions. Gauss's work was slow going because he did not blindly accept the published values but insisted on double-checking them, an exercise that paid off handsomely because he caught numerous errors. "Since I lacked the patience to go through the whole series systematically," he wrote to the astronomer Johann Franz Encke, "I have often used a spare quarter of an hour to investigate a thousand numbers here and there; at last I gave it up altogether, without ever finishing the first million. . . . The thousand numbers between 101,000 and 102,000 bristle with errors in Lambert's supplement; in my copies I have crossed out seven numbers that are not primes, and in return put in two that were missing." The hard work led Gauss to put forward the first formula that predicted the distribution of primes. Although the formula matched the long list of primes, Gauss never proved that his Prime Number Theorem was universally true. A full century would pass before Jacques Hadamard and Charles de là Vallée Poussin produced a proof.

Like Chebyshev's proof, the 1896 proof of the Prime Number Theorem depended on heavy machinery, and the brightest mathematical minds were convinced that the theorem couldn't be proved with anything less. Erdős and Atle Selberg, a colleague who was not yet well known, stunned the mathematics world with an "elementary" proof. According to Erdős's friends, the two agreed that they'd publish back-to-back papers in a leading journal delineating their respective contributions to the proof. Erdős then sent out postcards to mathematicians informing them that he and Selberg had conquered the Prime Number Theorem. Selberg apparently ran into a mathematician he didn't know who had received a postcard, and the mathematician immediately said, "Have you heard? Erdős and What's-His-Name have an elementary proof of the Prime Number Theorem." Reportedly, Selberg was so injured that he raced ahead and published without Erdős, and thus got the lion's share of credit for the proof. In 1950, Selberg alone was awarded the Fields Medal, the closest equivalent in mathematics to a Nobel Prize, in large part for his work on the Prime Number Theorem.

Priority fights are not uncommon in mathematics. Unlike other scientists, mathematicians leave no trail of laboratory results to substantiate who did what. Indeed, Erdős spent much time mediating priority fights among his closest collaborators. "When I was a graduate student," Joel Spencer said, "I thought only third-rate mathematicians would have these fights. But it's actually first-rate mathematicians. They're the ones who are passionate about mathematics." If they can't fathom what's in the SF's Book, they don't want anyone else to. The late R. L. Moore, a strong Texas mathematician, put it bluntly: "I'd rather a theorem not be thought of than I not be the one who thinks of it."

Erdős was singularly generous when it came to sharing mathematical ideas with his colleagues. "He would share his conjectures because his goal was not to be the first to prove something," said Alexander Soifer, who wrote two papers with him. "Rather his goal was to see that *somebody* proved it—with him or without him. Nobody was the wandering Jew as Paul was. He went around the world distributing his conjectures, his insights, to other mathematicians."

"Erdős contributed an enormous amount to mathematics," recalled Richard Guy, a number theorist at the University of Calgary. "But for me his even greater importance is that he created a large number of mathematicians. He was the problem poser *par excellence*. His ability to formulate problems of any level of difficulty is legendary. Many people can ask questions which are impossibly difficult or trivially easy. It is given to few of us to tread the narrow path between triviality and unattainability. Erdős problems were not Hilbert problems, which took half a century or more to settle. Erdős questions were always just right. So often, when we are fumbling with our research, it is because we are not asking the right question. Many of Erdős's questions have remained as outstanding but important problems, but most have been attacked and partially, or perhaps completely, solved.

"But Erdős not only asked the right question. He asked them of the right person. He knew better than you yourself knew what you were capable of. How many people must have got started on research by solving a $5.00, or maybe even a $1.00, Erdős problem? He gave the confidence that many of us needed to embark on mathematical research."

"He had the unique ability to identify problems which were just beyond what you could currently do," said Graham. "It's easy to state impossible problems but he would state a problem whereby if you could solve this problem you would then know more than you knew before and open the door wider. It was like climbing a mountain and being able to drive one more piton in to be able to work your way up the mountain."

Only one mathematician in history managed to publish more pages of mathematics than Erdős did. In the eighteenth century, the Swiss wizard Leonhard Euler, who fathered thirteen children, wrote eighty volumes of mathematical results, many reportedly penned in the thirty minutes between the first and second calls to dinner. Erdős, though, set a record for coming up with good problems and seeing that *somebody* solved them.

■

For Erdős, mathematics was a glorious combination of science and art. On the one hand, it was the science of certainty, because its conclusions were logically unassailable.

Unlike biologists, chemists, or even physicists, Erdős, Graham, and their fellow mathematicians *prove* things. Their conclusions follow syllogistically from premises, in the same way that the conclusion "Bill Clinton is mortal" follows from the premises "All presidents are mortal" and "Bill Clinton is a president." On the other hand, mathematics has an aesthetic side. A conjecture can be "obvious" or "unexpected." A result can be "trivial" or "beautiful." A proof can be "messy," "surprising," or, as Erdős would say, "straight from the Book." In a good proof, wrote Hardy, "there is a very high degree of *unexpectedness*, combined with *inevitability* and *economy*. The argument takes so odd and surprising a form; the weapons used seem so childishly simple when compared with the far-reaching consequences; but there is no escape from the conclusions."

What is more, a proof should ideally provide insight into why a particular result is true. Consider one of the most famous results in modern mathematics, the Four Color Map Theorem, which states that no more than four colors are needed to paint any conceivable flat map of real or imaginary countries in such a way that no two bordering countries have the same color. From the middle of the nineteenth century, most mathematicians believed that this seductively simple theorem was true, but for 124 years a parade of distinguished mathematicians and dedicated amateurs searched in vain for a proof—and a few contrarians looked for a counterexample. "When I started at AT&T," said Graham, "there was a mathematician there named E. F. Moore who was convinced that he could find a counterexample. Each day he would bring in a giant sheet of paper, and I mean giant, two feet by three feet, on which he had drawn a map with a few thousand countries. 'I know this one will require five colors,' he'd confidently announce in the morning and volunteer to give me a dollar if it wasn't the long-sought-after counterexample. Then he'd go off and spend hours coloring it. He'd come by at the end of the day, shake his head, and hand me a dollar. The next day he'd be back with another map and we'd go through the same thing again. It was the easiest way to make a buck!"

By 1976 it was clear why Moore's quest for a five-color map had come to nought. That was the year Kenneth Appel and Wolfgang Haken of the University of Illinois finally conquered this mathematical Mount Everest. When word of the proof of the Four Color Map Theorem reached college mathematics departments, instructors cut short their lectures and broke out champagne. Some days later they learned to their dismay that Appel and Haken's proof had made unprecedented use of high-speed computers: more than 1,000 hours logged among three machines. What Appel and Haken had done was to demonstrate that all possible maps were variations of more than 1,500 fundamental cases, each of which the computers was then able to paint using at most four colors. The proof was simply too long to be checked by hand, and some mathematicians feared that the computer might have slipped up and made a subtle error. Today, more than two decades later, validity of the proof is generally acknowledged, but many still regard it as unsatisfactory. "I'm not an expert on the four-color problem," Erdős said, "but I assume the proof is true. However, it's not beautiful. I'd prefer to see a proof that gives insight into why four colors are sufficient."

*Beauty* and *insight*—these are words that Erdős and his colleagues use freely but have difficulty explaining. "It's like asking why Beethoven's Ninth Symphony is beautiful," Erdős said. "If you don't see why, someone can't tell you. I know numbers are beautiful. If they aren't beautiful, nothing is."

Pythagoras of Samos evidently felt the same way. In the sixth century B.C. he made a kind of religion out of numbers, believing that they were not merely instruments of enumeration but friendly, perfect, sacred, lucky, or evil. He worshipped whole numbers and simple fractions, the ratios of whole numbers. And he took his religion seriously. When one of his followers challenged his world view by discovering that a common measure, namely, the length of the diagonal of a unit square (the square root of 2, or $\sqrt{2}$), could be expressed neither as a whole number nor as a simple fraction, Pythagoras became distraught and swore his disciples to secrecy.

Legend has it that when one of his followers subsequently betrayed him, Pythagoras had him executed—what Erdős called the "Pythagorean scandal."

Pythagoras certainly had his eccentricities—he was a vegetarian who refused to eat beans because they reminded him of testicles—but he got mathematics off to a solid footing by championing the concept of proof. He also had an uncanny feel for individual numbers. He considered 220 and 284 to be friendly. His notion of a "friendly" number was based on the idea that a human friend is a kind of alter ego. Pythagoras wrote, "[A friend] is the other I, such as are 220 and 284." These numbers have a special mathematical property: each is equal to the sum of the other's *proper divisors* (divisors other than the number itself). The proper divisors of 220 are 1, 2, 4, 5, 10, 11, 20, 22, 44, 55, and 110, and they sum to 284; the proper divisors of 284 are 1, 2, 4, 71, and 142, and they sum to 220.

A second pair of friendly numbers (17,296 and 18,416) was not discovered until 1636, by Pierre de Fermat. By the middle of the nineteenth century many able mathematicians had searched for pairs of friendly numbers, and some sixty had been found. But not until 1866 was the second *smallest* pair (1,184 and 1,210) discovered by a sixteen-year-old Italian schoolboy. By now hundreds of friendly numbers have been discovered, but, as with twin primes, even today no one knows whether there are infinitely many. Erdős thought their supply was inexhaustible, and he wrote one of the earliest papers in the literature on the distribution of friendly numbers. Why it should be so much easier to prove that the number of primes is infinite than it is to prove that the number of friendly numbers is infinite is one of the great unanswered meta-questions of mathematics.

Pythagoras saw perfection in any integer that equaled the sum of all the *other* integers that divided evenly into it. The first perfect number is 6. It's evenly divisible by 1, 2, and 3, and it's also the sum of 1, 2, and 3. The second perfect number is 28. Its divisors are 1, 2, 4, 7, and 14, and they add up to 28. During the Middle Ages, religious scholars asserted that the perfection of 6 and 28 was part of the fabric of the universe: God created the world in six days and the Moon orbits the Earth every twenty-eight days. St. Augustine believed that the properties of the numbers themselves, not any connection to the empirical world, made them perfect: "Six is a number perfect in itself, and not because God created all things in six days; rather that the inverse is true; God created all things in six days because this number is perfect. And it would remain perfect even if the work of the six days did not exist."

The ancient Greeks knew of only two perfect numbers besides 6 and 28: namely, 496 and 8,128. (Seventeen centuries would pass before the discovery of the fifth perfect number, 33,550,336.) Since the four perfect numbers the Greeks knew were all even, they wondered whether an odd perfect number exists. As of April 1998, mathematicians knew of thirty-seven perfect numbers, the largest having 1,819,050 digits, and all thirty-seven were even. Each time a new Mersenne prime $2^n - 1$ is discovered, a new even perfect number can be generated by multiplying $2^{n-1}$ by it. Thus the largest known prime, $2^{3,021,377} - 1$, yields the thirty-seventh even perfect number, $2^{3,021,376} (2^{3,021,377} - 1)$. Mathematicians cannot rule out the possibility that the thirty-eighth perfect number will be odd. Whether an odd perfect number exists is among the oldest unsolved problems in mathematics. Equally daunting is the unsolved problem of how many perfect numbers there are. No one knows whether they are finite or infinite in number.

When it comes to prime numbers and perfect numbers, said Erdős, "babies can ask questions that grown men can't answer." And Erdős was willing to discuss number theory with children and adults alike, with anyone who had an idea (and some who didn't). So open was Erdős to talking mathematics with just about anyone that his friends joked he couldn't ride on a train without proving a theorem with the conductor.

In 1985, a high school senior in Hawaii named David Williamson, who was taking mathematics courses at the University of Honolulu, proved that if an odd perfect number exists, it must have exactly one prime factor that, when divided by 4, leaves a remainder of 1. Williamson's professor didn't know whether the result was original and suggested he write to the legendary Erdős, who had just come through town. Williamson eventually got a letter back from Erdős: "The result you proved is in fact due to Euler. He also proved that every even perfect number is of the form $2^{p-1}$ $(2^p - 1)$ where $2^p - 1$ is a prime. It is also known (proved by Carl Pomerance) that an odd perfect number if it exists must have at least 7 distinct prime factors. Perhaps the following problem of mine will interest you. . . ." Williamson, now a combinatorialist at IBM, was thrilled: "This letter to a high school student won't rank very high on Erdős's list of accomplishments, but it did mean a lot to me."

Perfect numbers and friendly numbers are among the areas of mathematics in which child prodigies tend to show their stuff. Like chess and music, such areas do not require much technical expertise. No child prodigies exist among historians or legal scholars, because years are needed to master those disciplines. A child can learn the rules of chess in a few minutes, and native ability takes over from there. So it is with areas of mathematics like these, which are aspects of elementary number theory (the study of the integers), graph theory, and combinatorics (problems involving the counting and classifying of objects). You can easily explain prime numbers, perfect numbers, and friendly numbers to a child, and he or she can start playing around with them and exploring their properties. Many areas of mathematics, however, require technical expertise, which is acquired over years of assimilating definitions and previous results. By the time mathematical prodigies mature and enter college, they usually have the patience to master these more technical areas—and often go on to make great discoveries in them. Erdős was an exception. He stuck chiefly to the areas of mathematics in which prodigies excel.

This is not to say that his mathematical interests were narrow. On the contrary, he opened up whole new areas of mathematics. But these areas typically required a minimum of technical knowledge and, consequently, are stimulating new generations of prodigies.

Erdős's forte was coming up with short, clever solutions. He solved problems not by grinding out pages of equations but by constructing succinct, insightful arguments. He was a mathematical wit, and his shrewdness often extended to problems outside his areas of specialty. "In 1976, we were having coffee in the mathematics lounge at Texas A & M," recalled George Purdy, a geometer who began working with Erdős in 1967. "There was a problem on the blackboard in functional analysis, a field Erdős knew nothing about. I happened to know that two analysts had just come up with a thirty-page solution to the problem and were very proud of it. Erdős looked up at the board and said, 'What's that? Is it a problem?' I said yes, and he went up to the board and squinted at the tersely written statement. He asked a few questions about what the symbols represented, and then he effortlessly wrote down a two-line solution. If that's not magic, what is?"

Erdős was the consummate problem solver. Most elderly mathematicians, if they're still going strong, are theory builders. They have stopped solving problems and are setting a general agenda for mathematical research, pointing to new or neglected areas that younger talent should pursue. Not Erdős. As long as problems remained to be solved, he'd be slugging it out in the trenches.

His style was to work on many problems at once with colleagues in far-flung locations. "Every day he called mathematicians all over the world," said Peter Winkler of AT&T. "He called me all the time. 'Is Professor Winkler there?' Even when my kids were very young, they knew immediately that it was Uncle Paul. He knew every mathematician's phone number, but I don't think he knew anyone's first name. I doubt if he would have recognized my first name even though I worked with him for twenty years. The only person he called by his first name was Tom Trotter, whom he called Bill."

When Erdős collaborated in person, he liked to do work with several people at the same time. "He went around the room, like a grandmaster playing simultaneous chess," said Bruce Rothschild, who wrote eight papers with Erdős. "It was very stimulating. You had some time to think while he worked with the others before he got back to you. And you had a chance to see the problems that others were working on." Erdős was allowed to think about many problems at once, but he expected his collaborators to focus on the problem at hand. "No illegal thinking," he'd say when he sensed their minds wandering.

Erdős's ability to think about disparate things simultaneously was legendary. Michael Golomb, who wrote a joint paper with Erdős in 1955, recalled a time in the 1940s when he came across Erdős playing chess with a local master named Nat Fine, "whom Erdős could beat only rarely, usually by psychological warfare. . . . I saw Nat with his head between his hands, deep in thought considering the next move, while Erdős seemed to be engrossed in studying a voluminous encyclopedia of medicine. . . . I asked him, 'What are you doing, Paul? Aren't you playing against Nat?' His answer was, 'Don't interrupt me . . . I am proving a theorem.'"

■

One of the areas of mathematics that Erdős pioneered is a philosophically appealing aspect of combinatorics called Ramsey theory. It is the area in which Graham's record-setting number comes into play. The idea behind Ramsey theory is that complete disorder is impossible. The appearance of disorder is really a matter of scale. Any mathematical "object" can be found if sought in a large enough universe. "In the TV series *Cosmos*, Carl Sagan appealed to Ramsey theory without knowing that's what he was doing," Graham said. "Sagan said people often look up and see, say, eight stars that are almost in a straight line. Since the stars are lined up, the temptation is to think they were artificially put there, as beacons for an interstellar trade route, perhaps. Well, Sagan said, if you look at a large enough group of stars, you can see almost anything you want. That's Ramsey theory in action."

In Sagan's example, the mathematician would want to know the smallest group of arbitrarily positioned stars that will always contain eight that are nearly lined up. In general, the Ramsey theorist seeks the smallest "universe" that's guaranteed to contain a certain object. Suppose the object is not eight stars in a row but two people of the same sex. In this case, the Ramsey theorist wants to know the smallest number of people that will always include two people of the same sex. Obviously, the answer is three. Or, as Graham's colleague Daniel Kleitman put it, "Of three ordinary people, two must have the same sex."

Ramsey theory takes its name from Frank Plumpton Ramsey, a brilliant student of Bertrand Russell, G. E. Moore, Ludwig Wittgenstein, and John Maynard Keynes, who might well have surpassed his teachers had he not died of jaundice in 1930, at the age of twenty-six. While his brother Michael pursued the transcendent reality that theology offers (he became the archbishop of Canterbury), Frank Ramsey, a spirited atheist, pursued the transcendent reality that mathematics offers. He also studied philosophy and economics, writing two papers on taxation and savings that were heralded by Keynes and are still widely cited in the economics literature. But it is eight pages of mathematics that have made him eponymous—eight pages that Erdős seized on and developed into a full-fledged branch of mathematics. Like all the problems Erdős worked on, Ramsey problems can be simply stated, although the solutions are often hard to come by.

The classic Ramsey problem can be phrased in terms of guests at a party. What is the minimum number of guests that need to be invited so that either at least three guests will all know each other or at least three will be mutual strangers? Mathematicians, as is their trademark, are careful to articulate their assumptions. Here they assume that the relation of knowing someone is symmetric: If Sally knows Billy, Billy knows Sally. With this assumption in mind, consider a party of six. Call one of the guests David. Now, since David knows or doesn't know each of the other five, he will either know at least three of them or not know at least three. Assume the former (the argument works the same way if we assume the latter). Now consider what relationships David's three acquaintances might have among themselves. If any two of the three are acquaintances, they and David will constitute three who know each other—and we have our quorum. That leaves only the possibility that David's three acquaintances are all strangers to one another—but that achieves the quorum too, for they constitute three guests who do not know each other. To understand why a party of five is not enough to guarantee either three people all of whom know each other or three people none of whom do, ponder the case of Michael, who knows two and only two people, each of whom knows a different one of the two people Michael doesn't know.

We have just written out a mathematical proof, perhaps not one from the Book, but a proof nonetheless. And the proof provides insight into why a party of six must include at least three mutual acquaintances or three mutual strangers. Another way to prove this is by brute force, listing all conceivable combinations of acquaintanceship among six people—32,768 such possibilities exist—and checking to see that each combination includes the desired relationship. This brute-force proof, however, would not provide insight. Moreover, it would be time-consuming. Combinatorics is often described as "the art of counting without counting." To solve this Ramsey problem, the combinatorialist wants to avoid "counting" all 32,768 possibilities.

Suppose we want not a threesome but a foursome who either all know each other or are all mutual strangers. How large must the party be? Erdős and Graham and their fellow Ramsey theorists have proved that 18 guests are necessary and enough. But raise the ante again, to a fivesome, and no one knows how many guests are required. The answer is known to lie between 43 and 49. That much has been known for two decades, and Graham suspects that the precise number won't be found for at least a hundred years. The case of a sixsome is even more daunting, with the answer known to lie between 102 and 165. The range grows wider still for higher numbers.

Erdős liked to tell the story of an evil spirit that can ask you anything it wants. If you answer incorrectly, it will destroy humanity. "Suppose," Erdős said, "it decides to ask you the Ramsey party problem for the case of a fivesome. Your best tactic, I think, is to get all the mathematicians in the world to drop what they're doing and work on the problem, the brute-force approach of trying all the specific cases"—of which there are more than 10 to the 200th power (the number 1 followed by 200 zeroes). "But if the spirit asks about a sixsome, your best survival strategy would be to attack the spirit before it attacks you. There are too many cases even for computers." The most complex party problem that has been solved with the aid of computers, 110 of them running in sync, is the case of the minimal guest list needed to guarantee a foursome of friends or a fivesome of strangers. In 1993 the answer was found to be 25.

Graham's record-setting number comes up in a similar problem. Take any number of people, list every possible committee that can be formed from them, and consider every possible pair of committees. Now assign each pair of committees to one of two groups. No matter how the assignment is made, the "object" Graham wants to guarantee is four committees in which all the pairs fall in the same group and all the people belong to an even number of committees. How many people are required to guarantee the presence of four such committees? Graham suspects that the answer is six, but all that he or anyone else has been able to prove is that four such committees will always exist if the number of people is equal to his record-setting number. This astonishing gap between what is suspected, based on observations of specific cases, and what is known shows how hard Ramsey theory is.

A simpler example of Ramsey theory, often posed in mathematics competitions, involves taking the first 101 integers and arranging them in any order you like. No matter how perverse the arrangement, you'll always be able to find eleven integers that form either an increasing sequence or a decreasing sequence. "You don't have to pick the integers consecutively," Graham said. "You can jump. You might pick the first one, then the nineteenth one, then the twenty-second one, then the thirty-eighth one—but they all have to be going up or going down. This doesn't have to be the case if you take only the first 100 integers. How would I order the first 100 integers so that you can't find the desired sequence of eleven? I'd start 91, 92, 93, all the way up to 100. Then 81, 82, 83, up to 90. Then 71, 72, 73, up to 80. You get the idea." The full order, for those who don't get the idea, is

91, 92, 93, 94, 95, 96, 97, 98, 99, 100, 81, 82, 83, 84, 85, 86, 87, 88, 89, 90, 71, 72, 73, 74, 75, 76, 77, 78, 79, 80, 61, 62, 63, 64, 65, 66, 67, 68, 69, 70, 51, 52, 53, 54, 55, 56, 57, 58, 59, 60, 41, 42, 43, 44, 45, 46, 47, 48, 49, 50, 31, 32, 33, 34, 35, 36, 37, 38, 39, 40, 21, 22, 23, 24, 25, 26, 27, 28, 29, 30, 11, 12, 13, 14, 15, 16, 17, 18, 19, 20, 1, 2, 3, 4, 5, 6, 7, 8, 9, 10.

The longest increasing sequence is a block of ten consecutive numbers, say, 81 through 90, or 41 through 50; no eleventh number lies to the right of either block. For a decreasing sequence, you can take any number from the 90s, then any number from the 80s, and so on for each block of 10 integers, but that gives you only 10 decreasing numbers. But add the number 101 anywhere you want— you can even rearrange the 101 numbers anyway you want—and you're bound to find an eleven-term sequence with all the numbers rising or all the numbers falling. Ramsey theory, says Graham, makes a generalization of this result: to guarantee either a rising or falling sequence of length $n + 1$, you need $n^2 + 1$ numbers; with $n^2$ numbers, you may not get it.

Graham, whose old license plate reads RAMSEY, thinks that centuries may pass before much of Erdős's and his work in Ramsey theory has significant applications in physics, engineering, or elsewhere in the real world, including his place of employment, AT&T. "The applications aren't the point," Graham said. "I look at mathematics pretty globally. It represents the ultimate structure and order. And I associate doing mathematics with control. Jugglers like to be able to control a situation. There's a well-known saying in juggling: 'The trouble is that the balls go where you throw them.' It's just you. It's not the phases of the moon, or someone else's fault. It's like chess. It's all out in the open. Mathematics is really there, for you to discover. The Prime Number Theorem was the same theorem before people were here, and it will be the same theorem after we're all gone. It's the Prime Number Theorem."

"In a way," Erdős said, "mathematics is the only infinite human activity. It is conceivable that humanity could eventually learn everything in physics or biology. But humanity certainly won't ever be able to find out everything in mathematics, because the subject is infinite. Numbers themselves are infinite. That's why mathematics is really my only interest." One can reconstruct chapters of the SF's Book, but only the SF has it from beginning to end.

"The trouble with the integers is that we have examined only the small ones," said Graham, an odd remark for a man who set a world record for looking at the largest integer ever studied. "Maybe all the exciting stuff happens at really big numbers, ones we can't get our hands on or even begin to think about in any very definite way. So maybe all the action is really inaccessible and we're just fiddling around. Our brains have evolved to get us out of the rain, find where the berries are, and keep us from getting killed. Our brains did not evolve to help us grasp really large numbers or to look at things in a hundred thousand dimensions. I've had this image of a creature, in another galaxy perhaps, a child creature, and he's playing a game with his friends. For a moment he's distracted. He just thinks about numbers, primes, a simple proof of the twin-prime conjecture, and much more. Then he loses interest and returns to his game."

We Earthlings, where are we in our understanding of numbers? Each result—say, Erdős's proof that a prime can always be found between an integer and its double—although touted in the mathematics journals, is only an imperceptible advance toward some kind of cosmic understanding of the integers. "It will be millions of years before we'll have any understanding," Erdős said, "and even then it won't be a complete understanding, because we're up against the infinite."

# 2

## EPSZI'S ENIGMA

*Temetni tudunk.*
(How to bury people—that is one thing we
know.)

—old Magyar saying

It was 1930, and fourteen-year-old Andrew Vázsonyi in Budapest was showing signs of mathematical talent. His father, who owned the best shoe store in town and knew everyone who was anyone—or at least knew their shoe size—wanted to find his son an intellectual playmate. Erdős's reputation as a math whiz was already well known—his photo had appeared in a famous high school mathematics magazine—and so Vázsonyi's father invited the seventeen-year-old Erdős to come play with Andrew at his store.

"I'm not sure why we were meeting at the store," recalled Vázsonyi sixty-seven years later. "I don't understand why I didn't go to his place or he didn't come to my home, although at the time we lived in a pretty miserable place. My father's store was very narrow and very deep. It was elegant mahogany, with boxes of shoes stacked to the ceiling on both sides. In the back he had a small office, with a steel moneybox, a desk, and a telephone. Erdős and I were supposed to meet there. I sat in the back waiting."

Erdős proved to be a "weirdo" (in the words of Kathy, the shoe store saleswoman) from the start. He knocked on the store door, which was no more the custom in Hungary than it was here. After that unusual civility, he dispensed with all introductions and conversational pleasantries and charged to the back.

"Give me a four-digit number," he demanded.

"2,532," Vázsonyi replied.

"The square of it is 6,411,024," Erdős said. "Sorry. I am getting old and cannot tell you the cube."

"How many proofs of the Pythagorean Theorem do you know?" Erdős asked.

"One," Vázsonyi said.

"I know thirty-seven. Did you know that the points of a line do not form a denumerable set?" Erdős proceeded to show Vázsonyi a proof and then announced he had to run.

"When Erdős said he must run," recalled Vázsonyi, the statement was literally true. He never walked but instead "kind of cantered down the street like a big ape, hunched over, moving sideways, his arms swinging. People always turned around and stared. We used to go skating all the time, and he skated like an ape, too. It was embarrassing because I was trying to meet girls on the ice. Of course, he wasn't interested in that. As he got older, his gait was less apelike, although still strange. He always moved fast. And he developed this thing about running up to a wall, suddenly stopping short, turning around abruptly, and running back. Once he didn't stop in time. He smashed into the wall and hurt himself.

"I'm not sure why he acted the way he did when we first met—all the stuff about squaring numbers and the various proofs. As I got to know him, I discovered he was not the kind of person who showed off. His whole outlook on life wasn't that way. So I can't really explain why he was talking about all the proofs he knew. Most of this stuff I couldn't understand. I discovered later that he was notorious for being incomprehensible." Erdős would leave out steps and assume that his listeners knew things that they didn't. What was clear to his rarefied mind was not necessarily clear to the mere mortals who surrounded him. Vázsonyi remembered being particularly confused by Erdős's proof about points on a line. "I knew Paul Turán at that time. He lived upstairs, in an apartment above my father's store. When Erdős left after that first encounter, I went up to see Turán, who said there was something incomplete about the proof Erdős had told me. Of course, it's possible that Turán was wrong and the proof was right.

"Erdős and I became friends pretty fast. Soon we saw each other almost daily. We went skating, we played Ping-Pong regularly at Turán's place, we went hiking, and most of all we talked math. There was all sorts of gossip about Erdős going around. My mother said that she heard bad news about Erdős—that he was homosexual, because of his strange attitude toward women, and stories that his own mother still dressed and bathed him. But the stuff about homosexuality was nonsense."

■

Erdős was born in Budapest on March 26, 1913, the son of two high school mathematics teachers. While his mother, Anna, was in the hospital giving birth to him, her two daughters, ages three and five, contracted septic scarlet fever and died within the day. "It was something my mother didn't like to talk about," Erdős said. "Their names were Klára and Magda, I think." Of the three children, the girls were considered to be the smart ones. "No mother could ever recover from such a loss," said Cousin Fredro. "She never did."

■

Death and tragedy have long been part of the Hungarian character. According to the historian John Lukacs, on All Souls' Day in Erdős's youth, "thousands of people streamed toward the cemeteries of Budapest, with flowers in their hands, on that holy day which is perhaps taken more seriously in Hungary than anywhere else because of the national temperament. *Temetni tudunk*—a terse Magyar phrase whose translation requires as many as ten English words to give its proper (and even then, not wholly exact) sense: 'How to bury people—that is one thing we know.' " The phrase was coined long before Hungary had experienced the devastation of the two world wars, whose carnage would come disproportionately from within its borders.

Few countries have had as violent a history as Hungary. In the ninth century, nomadic Magyar warriors from the steppes of Eastern Europe crossed the Carpathian Mountains and, renouncing their peripatetic way of life, settled in the middle Danube basin at the heart of what is now Hungary. The Magyars, also known as the On-Ogurs ("people of the ten arrows"), were skilled archers and javelin throwers who raided their neighbors, plundering Germany from Bavaria to Saxony. Recent converts to Christianity, the Magyars had to defend their own new territory against a series of invaders. They held their own until 1241, when several hundred thousand Mongol horsemen from Genghis Khan's empire slaughtered half the Magyar population and enslaved much of the rest, a bloodbath from which the country barely recovered. "The problem with Hungary,"

Erdős once joked, reflecting on his country's history, "is that in every war we chose the wrong side."

A succession of foreign kings ruled Hungary and inadequately defended the country against repeated Turkish incursions. At the Battle of Mohács on August 29, 1526, the Turkish army of Sultan Süleyman I slew the Magyar king, destroyed his troops, and celebrated with an extended looting spree that left 200,000 people dead, another 100,000 enslaved, and much of the country's gold, silver, and jewels missing. Whole villages were wiped out, forests were torched, and farmland leveled. In 1541, the Ottoman overlords destroyed the historic Kingdom of Hungary by trisecting the Magyar territory, ceding its northern and western regions to the Hapsburgs, making Transylvania an independent Muslim state, and keeping the twin cities of Buda and Pest for themselves. In 1699, the Turks finally withdrew from the twin cities, and the Hapsburgs expanded their authority there. In 1867, Hungary got back some if its independence with the creation of the Austro-Hungarian dual monarchy, in which Hungary was granted a separate parliament and semi-autonomy, although it was still part of the Hapsburg kingdom of Franz Joseph of Austria. In 1873, the palace hill town of Buda and the reclaimed marshlands of Pest, on opposite banks of the Danube, officially became one city.

The Budapest of Erdős's birth was a sophisticated modern city, home of the largest stock exchange in Europe and the most grandiose parliament building in the world. Budapest rivaled Paris and Vienna in first-class hotels, garden restaurants, and late-night cafés, which were hothouses of "illicit trading, adultery, puns, gossip and poetry, the meeting places for the intellectuals and those opposed to oppression." It was also an Old World city, its women praised for their beauty and its men for their chivalry. Budapest "has never been an agreeable town. But desirable, yes: like a racy, full-blooded young married woman about whose flirtations everyone knows and yet gentlemen are glad to bend down and kiss her hand," wrote Gyula Krúdy, an acclaimed writer of Magyar prose, who moved from the Hungarian provinces to Budapest in 1896, the year the city's subway system opened, the first in Europe. Krúdy's enthusiasm for Budapest was irrepressible:

> Here the dancing in the theaters is the best, here everyone in a crowd may think that he is a gentleman even if he had left jail the day before; the physicians' cures are wonderful, the lawyers are world-famous, even the renter of the smallest rooms has his bath, the shopkeepers are inventive, the policemen guard the public peace, the gentlefolk are agreeable, the streetlights burn till the morning, the janitors will not allow a single ghost inside, the tramcars will carry you to the farthest places within an hour, the city clerks look down on the state employees, the women are well-read from their theater magazines, the porters greet you humbly on the street corners, the innkeeper inquires of your appetite with his hat in hand, the coach drivers wait for you solemnly during an entire day, the salesgirls swear that your wife is the most beautiful of women, other girls in the night clubs and orpheums hear out your political opinions politely, you find yourself praised in the morning newspaper after you had witnessed an accident, well-known men use the spittoons in the café gardens, you are being helped into your overcoat, and the undertaker shows his thirty-two gold teeth when you take your leave from this city forever.

Before World War I, life for Jewish families like Erdős's and the Vázsonyis was more comfortable in Budapest than in most other places in Europe. The Hapsburgs had emancipated Hungary's Jews in 1867, and Jewish immigration was encouraged by the Magyar ruling class for the simple reason that the Magyars were numerically in the minority and assimilated Jews added to their ranks. To encourage assimilation, the Hapsburgs granted royal titles to prominent Jews. For instance, the banker Max von Neumann, father of the mathematician John von Neumann, was ennobled in 1913. "The Hapsburgs were not really anti-Semitic," said Vázsonyi, "which meant in Hungary that they didn't dislike the Jews any more than was proper." Vázsonyi's family was very prominent under the Hapsburgs, and one of his uncles was in the cabinet of Charles IV, who was crowned in 1916. Many assimilated Jewish families took on Hungarian names. The Vázsonyi name was originally Weiszfeld, and the Erdős name, which means *wooded* in Hungarian, was Engländer. Erdős's parents were not observant Jews, though his grandparents were. According to László Babai, who wrote three papers with Erdős, when Erdős's father was courting his mother, he visited her once on Yom Kippur, a holy day of introspection, and "found her fasting and reading Maupassant. He pointed out the paradox he sensed in this combination. She agreed. Shedding tears, Anna relinquished the old traditions."

Through Magyarization, and the opportunities opened to them as card-carrying Hungarians, many Jewish families became economically well off. In 1910, Jews made up only 5 percent of Hungary's population, but they owned 37.5 percent of the farmland and constituted 51 percent of all lawyers, 60 percent of doctors, and 80 percent of financiers. Hungary, however, was not a Jewish paradise. Resentment over Jewish prosperity was building, but before anti-Semitism could really flare up, the events of June 28, 1914, distracted the entire country, rallying it around a common enemy. On a state visit to Bosnia, Archduke Franz Ferdinand, heir to the Hapsburg throne, was assassinated by a Serbian patriot, Gavrillo Princip. Austria-Hungary immediately declared war on Serbia, Russia declared war on Austria-Hungary, and all of Europe was plunged into World War I. In August, when Erdős was one and a half, his father, Lajos, was captured in a Russian offensive and sent to Siberia for six years.

With his father in prison and his mother teaching school during the day, Erdős was raised by a German governess. He became proficient with numbers as a toddler by studying the calendar and figuring out how many days it would be before his mother was home for the holidays. At four he entertained himself by "computing crazy things like how long it would take a train to reach the Sun." He amused his mother's friends by asking them how old they were and then calculating in his head how many seconds they had lived. He knew then that he wanted to be a mathematician, although he would pay attention to subsequent tutorials in history, politics, and biology. As soon as he could read, his mother plied him with medical literature, which he eagerly studied. She apparently had vague hopes that he might become a doctor. For most of Erdős's childhood his mother kept him out of school, fearing that it was the source of deadly childhood contagions. He studied at home with a tutor. Erdős stayed home until high school, and even then he went only every other year, because his mother kept changing her mind.

When the war wound down in 1918, a defeated Hungary was thrown into chaos. Some 734,000 prisoners of war, held mostly in Russia, were unaccounted for. The death toll of prisoners would reach 431,000, and Erdős and his mother didn't know if his father was among the dead. The Austro-Hungarian monarchy was dissolved on October 31, in the bloodless Aster Revolution. The monarchy's troops, to signal their support for democratic reform, ripped off their royal emblems and adorned themselves with red and white asters, available throughout the city in celebration of All Souls' Day. But the democratic Republic of Hungary, proclaimed with such promise and fanfare on November 16, lasted only four months. In the absence of a formal peace treaty ending World War I, armed marauders from Romania, Czechoslovakia, and Yugoslavia continued to ravage the Hungarian countryside, and the new democratic government was powerless to stop them.

On March 21, 1919, Béla Kun, a Transylvanian Jew who as a Russian prisoner of war had become a devotee of Lenin, staged a nonviolent coup that overnight transformed the Republic of Hungary into the Hungarian Soviet Republic. "The rousing tunes of the *Marseillaise* and of the *Internationale* drowned the music-loving town on the Danube in a fiery, melodious flood," Arthur Koestler wrote in his autobiography, *Arrow in the Blue*. "As Vienna had danced to the fiddle of Johann Strauss, so the people of Budapest now marched to the tune of the *Marseillaise*." But Kun's regime didn't even last as long as his predecessor's. His ministers were inept—it was said that his finance commissioner didn't know how to endorse a check. Budapest was slowly starving, and Kun did nothing to alleviate the food shortage. "The people of Budapest seemed to live mainly on ice-cream," Koestler wrote. "The only things ration cards and the paper money issued by the red regime would buy were cabbages, frozen turnips—and ice-cream .... It may be suspected my sympathies for the Commune were influenced by the fantastic amount of [ice cream] we ate, for breakfast, lunch, and dinner, during the hundred days. This suspicion would be the most unjust as there was only the one kind, vanilla, which I dislike."

Kun's soldiers, while keeping their guns silent, billeted themselves in the apartments of ordinary citizens. Koestler's mother managed to expel two soldiers. But the family of eleven-year-old Edward Teller, who would grow up to rail against Communism and build the hydrogen bomb, was not as lucky. Teller's family had to put up with two soldiers who slept on their couches and urinated on the rubber plant. Teller recalled hearing stories of Kun's opponents being hung from lampposts, but never saw any corpses himself. Five hundred people died at the hands of Lenin Boys, an undisciplined gang of hooligans, ex-military men, and escaped Soviet convicts who harassed and tortured bourgeois politicians, businessmen, and Catholic priests, whose churches Lenin Boys desecrated. Teller's mother, Ilona, who was Jewish, told their governess: "I shiver at what my people are doing. When this is over there will be a terrible revenge."

Kun's support collapsed in the countryside once he nationalized the farmland instead of following through on his promise to redistribute it to the peasantry. When help from Lenin never materialized, Kun watched impotently as Romanian troops advanced further into Hungary, coming within fifty miles of Budapest by the end of July 1919. In power only 133 days, Kun fled to Vienna (and eventually to Russia, where in 1937 Stalin had him shot as a traitor). Kun was replaced by Miklós Horthy, a former admiral of the Austro-Hungarian Navy, who rode in on a white horse and installed Europe's first postwar Fascist regime, reproaching Budapest as "the guilty city." He would serve as Hungary's dictator until 1945. Strong as his rule would become, he could not stop the victors of World War I from imposing a punitive peace treaty. On June 4, 1920, the Treaty of Trianon dismembered Hungary, stripping away 68 percent of its land and 59 percent of its population. Of the 10 million people who spoke Magyar, fully a third found themselves living in territory annexed by Romania, Czechoslovakia, or Yugoslavia.

During Kun's brief reign, Erdős's mother was promoted to principal of her high school. When right-wing groups called for a general strike against the Communist government, she stayed at work, out of deference not to Kun but to the children whose education would be disrupted without her. That decision cost Anna Erdős her job. Once Horthy took over, she was summarily dismissed, and couldn't teach again in a public school for twenty-six years, until the Communists finally returned to power.

No sooner had Horthy arrived than he unleashed a wave of White Terror, ten times deadlier and far more organized than the Red Terror of a few days before. Ilona Teller's fears came true. Armed Fascist thugs intent on ridding Budapest of Communist sympathizers targeted Jews in particular, who were assumed to be in league with their coreligionist Kun. Five thousand people were killed, many others were tortured, and tens of thousands of Jews were driven into exile. Edward Teller, John von Neumann, and the physicists Leo Szilard and Eugene Wigner all left Hungary for Germany, "then the mecca of scientists, only to be driven away by Hitler to eventual sanctuary in the United States. To this extent, Béla Kun can unwittingly take credit for the American preeminence in the development of nuclear energy": all four men lent their scientific creativity to making the atomic bomb.

Anna Erdős stayed in Budapest and feared for the safety of her son. From the balcony of their apartment they could see Jews being beaten in the streets. (Before Wigner left the country, he was attacked by a mob.) "There were a lot of anti-Semitic acts," Paul Erdős recalled. "Being a Jew, my mother once said to me, 'You know the Jews have such a difficult time, shouldn't we get baptized?' I told my mother, 'Well, you can do what you please, but I remain what I was born.' It was very remarkable for a small child—I was only six or seven then—because, actually, being Jewish meant nothing to me. It never did." But what meant everything to Erdős, even at this young age, was being true to his own birthright and never compromising his principles, no matter how inconvenient or life threatening it was to maintain them. Throughout his life he would fearlessly defy "Fascist" authorities of every stripe, be they armed thugs, mindless university bureaucrats, the U.S. Immigration Service, the Hungarian secret police, the FBI, Los Angeles traffic cops, or the SF Himself.

Although Erdős hated Fascists, he loved the word and applied it liberally to anything he didn't like. "Paul and I once went to a colleague's apartment where some kittens had been born," recalled Melvin Henriksen, who wrote one joint paper with Erdős. "Paul picked one up, but returned it gently to the box when the kitten scratched. 'Fascist cat!' he exclaimed. The lady of the house questioned how he could call such a little kitten a Fascist. 'If you were a mouse,' he said, 'you'd know.'"

Erdős and his mother survived the White Terror, and in 1920 his father returned home, haggard but happy. "*Apuka* [Daddy]!" Erdős yelled. "You look really old!" To kill time in the prison camp, Lajos had taught himself English from a book. Now he'd teach it to his son. "That's why my accent is so bad," Erdős said, "because I learned English from someone who had never heard it from a native speaker." (And Erdős's English didn't get much better once he lived in Britain and America. "I remember when

I first heard him speak," said John Selfridge, who wrote 13
papers with Erdős. "It was in 1946. I was a freshman in
Seattle, and I went down to Stanford when I learned he
was giving a talk. I was all excited, but my ear was so
badly tuned to Erdős's accent that I couldn't understand a
word except when he said, 'It's a little warm in here.
Shouldn't I open the window?' My friends told me he
talked about some pretty interesting problems." An American documentary about Erdős felt compelled to add sub-
titles to his garbled English. "Vot vuz zat ven it vuz
live?"—What was that when it was alive?—he always
asked when served a piece of meat.)

The White Terror had succeeded in silencing potential
opposition to the Fascist regime, and so Horthy called off
the violence but instituted laws to keep the Jews in their
place. In 1920, the notorious *Numerus Clausus*—the first
major anti-Jewish legislation in postwar Europe—limited
Jewish admissions to universities to 6 percent, which was
the percentage Jews represented in the overall population.
By the age of twelve, Erdős had determined that he would
eventually have to leave Hungary on account of his religion. But in 1928 the admissions restrictions were eased to
allow the winners of national competitions, no matter what
their religion, to go straight to college without taking entrance exams. That's how Erdős, at the age of seventeen,
ended up enrolling in 1930 in the University Pázmány
Péter in Budapest, where he graduated four years later with
a Ph.D. in mathematics.

With anti-Semitism on the rise in the early 1930s,
Erdős, Vázsonyi, and a few other young Jewish mathematicians got together weekly to exchange gossip, talk politics,
and, most important, do mathematics. On Sundays they'd
go hiking in the hills outside Budapest. "These were long
hikes," said Vázsonyi. "We used to meet at Berlin Place

and take the trolley to Zugliget. There were usually four or five of us, but sometimes as many as twenty." The young mathematicians also liked to meet in the city park at a bench by the hooded bronze statue of *Anonymous*, a historian who chronicled twelfth-century Hungarian kings. "When we got together as a group," Vázsonyi recalled, "we were always concerned that the police would come and question us. Group meetings were prohibited during the Horthy dictatorship. We could not speak freely. We thought there were spies everywhere. That's when Erdős started developing his private language. Many of us were Communists in the sense of what it meant at that time: that we were against the Horthy regime." But it wasn't safe to use the word *Communist* out loud, so Erdős started referring to Communists as people "on the long wavelength," because in the electromagnetic spectrum the red waves were long. He said that Horthy supporters and other Fascist sympathizers were "on the short wavelength." That's also when he started calling children and other small things "epsilons," grandchildren "epsilons squared," alcohol "poison," music "noise," and women "bosses," an inversion of what Hungarian women often called their husbands. "Give me an epsilon of poison," Erdős would say when he wanted a sip of wine. "Wine, women, and song" became "Poison, bosses, and noise." In his world, though, all children were actually bosses, regardless of their gender. "Women are bosses and men slaves but children are bosses *per se*," said Erdős. "So I was asked once: 'When does a slave child become a slave? If he is a boss originally, when does he become a slave?' And here I answered immediately: 'When he starts running after bosses.' "

"The language was very contagious," recalled Vázsonyi, "and continued to spread throughout the universal club of mathematicians" all over the world long after the Horthy regime was gone. Over time Erdős added new words such as "Joe" and "Joedom" for the USSR and "Sam" and "Samland" for the United States. International news, which he pronounced *nevsh*, became "the Sam and Joe show," and he invented a top-secret police organization called the FBU, a combined unit of the FBI and OGPU (the Soviet secret police agency that predated the KGB), which he said swapped agents back and forth between Joe and Sam. In the 1940s, he started calling God the SF. "With so many bad things in the world," he said, "I'm not sure that God, should He exist, is good." Erdős spoke approvingly of Anatole France's novel *The Revolt of the Angels*, in which God was portrayed as evil and the devil as benign.

Erdős didn't hesitate to use his special language outside mathematical circles. For instance, he once asked Barbara Piranian, president of the League of Women Voters in Ann Arbor, Michigan, "When will you bosses take the vote away from the slaves?" (And she answered, "There is no need; we tell them how to vote anyway.") But not all outsiders reacted to Erdősese with as good humor as Piranian. Cedric Smith, a statistician at University College London, recalled how soon after he had gotten married, Erdős mistook his mother-in-law for his wife and she "complained of the phone ringing and a deep voice (guess who?) saying, 'How are you? Vair is your slave? Is he preaching?' (in conventional English . . . Where is your husband? Is he lecturing?). Said she, indignantly, 'I haven't got a slave.' "

At the foot of *Anonymous*, discussions in Erdősese about politics and family always gave way to talk about math. "We were addicted to math," said Vázsonyi, and Erdős was the biggest junkie of all. "He lived in the world of prime numbers. They were his world. He had some peculiar affinity to prime numbers. When he was twenty, he announced he was going to prove the great Prime Number Theorem by elementary means using the ancient sieve of Eratosthenes. And sure enough, twenty years later, he'd do that. Prime numbers were on his mind all his life."

In late 1932 the *Anonymous* group included Esther Klein, a talented student back from a semester away at the University at Göttingen, and György (George) Szekeres, a recent chemistry graduate, eager to put aside his test tubes for mathematics. "Paul, then still a young student but already with a few victories in his bag, was always full of problems and his sayings were already a legend," recalled Szekeres. "He used to address us in the same fashion as we would sign our names under an article and this habit became universal among us; even today I often call old members of the circle by a distortion of their initials. 'Szekeres Gy, open up your wise mind.' This was Paul's customary invitation—or was it an order?—to listen to a proof or a problem of his."

On one of their weekly sojourns it was Klein who shared with the men a curious problem in plane geometry: given any five points on a flat surface, with no three of them in a straight line, prove that four of these points will always form a convex quadrilateral. (A *quadrilateral* is the generic term for any four-sided figure; squares, rectangles, and parallelograms are all quadrilaterals. By "convex," she meant that any spot within the quadrilateral has a direct line of sight to any other spot; a square, therefore, is a convex quadrilateral, whereas the shape of an arrowhead is not, because a spot in one of the back tips cannot "see" a spot in the other back tip. Another way of characterizing a convex figure is to say that it has no indentations.) Klein was able to prove the theorem by showing that all conceivable arrangements of five points fell into three general cases, each of which guaranteed a convex quadrilateral.

The first case is when the five points themselves form a convex pentagon; then any four of the five points form a convex quadrilateral:

The second case is when one of the five points lies inside the other four; in that case the four exterior points form the convex quadrilateral:

The remaining case is when two of the points lie inside the triangle formed by the other three. If a line is drawn through those two points, bifurcating the triangle, two points of the triangle will fall on one side of the line. Those four points always form a convex quadrilateral:

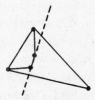

Erdős and Szekeres were taken with the elegant proof, so they tried to extend the result to polygons with many more sides. Another mathematician in their group, Endre Makai, soon proved that nine points were needed to guarantee a convex pentagon. Eight points, a counterexample showed, was one point too few:

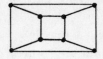

From such specific examples a strong generalization quickly emerged: was a convex polygon of $n$ sides guaranteed whenever the points numbered $2^{n-2} + 1$? The formula $2^{n-2} + 1$ obviously worked for a convex quadrilateral: when $n = 4$, $2^2 + 1$ is 5. Similarly, for $n = 5$, the case of the convex pentagon, the formula correctly gave $2^3 + 1$, or 9 points. But a proof of the generalization eluded the three Hungarians.

"We soon realized," recalled Szekeres, "that a simpleminded argument would not do and there was a feeling of excitement that a new type of geometrical problem emerged from our circle which we were only too eager to solve. For me that it came from Epszi, Paul's nickname for Esther, short for epsilon, added a strong incentive to be the first with a solution and after a few weeks I was able to confront Paul with a triumphant 'E.P., open up your wise mind.'" Szekeres proved the necessity of convex polygons of any given size $n$, but the proof gave, as the numbers of points required "an absurdly large value, nowhere near the suspected" $2^{n-2} + 1$. Nonetheless, Szekeres's accomplishment was impressive enough to win Klein's hand, and the two were married four years later. As a result, Erdős dubbed the problem the Happy End Problem, and that name has stuck in the mathematical literature.

Good thing Erdős wasn't a ladies' man, though, because he soon improved on Szekeres's result, although the gap between the suspected value and Erdős's proven value was still huge. Erdős proved that 71 points would always guarantee a convex hexagon, although 17 points ($2^4 + 1$ points) were thought to be sufficient. This was Erdős's first Ramsey result—in a large enough set of points, convex polygons cannot be avoided—although at the time neither he nor Klein or Szekeres had ever heard of Ramsey. Sixty years later, Klein and Szekeres are still married. They live in Sydney, Australia (the country Erdős called "Ned," after Ned Kelley, the daring Australian bandit of the nineteenth century), where they continue to prove and conjecture, although not at the frenetic pace of the *Anonymous* years.

There's something about the Happy End Problem that succumbs to couples in love. No progress at all was made on the problem until November 1996, when Graham and his wife decided to tackle it. "We had just seen Szekeres at Erdős's memorial service and we talked about how someone should make a dent in the gap after sixty years," Graham explained. "Then Fan and I were flying to New Zealand for a math conference. The flight takes a day, so we passed the time by working to narrow the gap. We succeeded in lowering the upper bound by one point. Okay, in absolute terms that's not much to get excited about. But it was exciting to make the first progress on the problem in sixty years—the problem is that intractable. To make the progress we did make called for some new ideas. Who knows? Someone else may look at these ideas and realize that if you do thus-and-such with them, you can make a bigger dent in the problem. That's how mathematical progress works. There's an old saying in mathematics:

> Problems worthy of attack,
> Prove their worth by fighting back.

This problem certainly fought back."

And just as Graham had hoped, as soon as their Air New Zealand success became known, other mathematicians were able to jigger the key ideas to further chip away at the upper bound. "It's kind of funny, this Klein-Erdős-Szekeres problem. Fan and I saved one point. So only 70 points were needed to guarantee a convex hexagon, with the suspected number still 17. Then Daniel Kleitman and Lior Pachter saved 5 points, reducing the guarantee to 65. After sixty years! Then yesterday," Graham said in the spring of 1997, "I received a paper where they cut the number of points in half. Whoom! They did it in a very nice way. Still, the gap is gigantic, the range is from 37 to 17, but we're plugging away at it. That's often how mathematical progress is made. Paul would be proud, because we did exactly what he'd want us to do. We had this problem that had gotten under our skin, and at first we couldn't do anything to it. Well, at least do something, Paul would urge. Crack it. Break it open a little bit. Then someone else will figure out how to break it open a little bit more."

■

In 1934, the year after the Happy End Problem was originally solved, Erdős left Hungary for political reasons. "I was Jewish," he said, "and Hungary was a semi-Fascist country." He went to England for a four-year postdoctoral fellowship at Manchester, where the wanderlust that would characterize his adult life was first evident. "From 1934 he hardly ever slept in the same bed for seven consecutive nights," wrote Erdős's collaborator Béla Bollabás, "frequently leaving Manchester for Cambridge, London, Bristol, and other universities." All along "he had intended to go to Germany, but the rise of Fascism prevented him. Jokingly he said that the traditional Jewish toast should be changed to 'Next year in Göttingen!' "

In Cambridge, on his second day in England, Erdős met

G. H. Hardy, the father of modern analytic number theory, who shared Erdős's dislike for authority and his irreverence toward God. Even as a child Hardy was a militant atheist, and as an adult he refused to set foot in houses of worship, even for such nonsectarian purposes as the traditional election of officers at his college. (To accommodate him, the college wrote a special exemption into its bylaws so that Hardy "could discharge certain duties by proxy which would otherwise have required him to attend Chapel.") Like Erdős, Hardy played games with the SF, even though he doubted His existence. On a turbulent boat trip from Scandinavia to England, Hardy dashed off a postcard to a colleague announcing that he had proved the Riemann Hypothesis, an important open conjecture in prime number theory. He figured the SF, whom he referred to as his "personal enemy," would not let him die with the potential of people thinking he had proved the Riemann Hypothesis. Although Hardy was dismissive of religion, he had—in the words of his obituary in *Oxford Magazine*—a "profound conviction that the truths of mathematics described a bright and clear universe, exquisite and beautiful in its structure, in comparison with which the physical world was turbid and confused. It was this which made his friends ... think that in his attitude to mathematics there was something which, being essentially spiritual, was near to religion."

If there were a Hall of Fame for eccentrics, Hardy and Erdős would be among the first inductees. Hardy was acutely self-conscious from a young age, and twice tried to take his own life. He deliberately put wrong answers down on exams in grade school so as to be spared the ordeal of coming in first and being awarded a prize in front of the class. He refused to wear a watch and avoided other mechanical objects from fountain pens to telephones. He could not stand to have his photograph taken or look at himself in the mirror, going so far as to shave by touch. When he checked into a hotel, the first thing he did was to cover the mirrors with towels. He thought he had the face of a gargoyle but in fact he was boyishly handsome, even in his fifties. Bertrand Russell praised his "bright eyes"; his colleague John Edensor Littlewood thought him "beautiful"; the Columbia University mathematician James Newman found him "strikingly handsome and graceful" with "his delicately cut but strong features, his high coloring, and the hair that fell in irregular bangs across his forehead"; and the novelist C. P. Snow approved of his tan, "a kind of Red Indian bronze. His face was beautiful— with high cheekbones, thin nose, spiritual and austere."

Hardy's first love was mathematics, and his second cricket. He peppered his mathematical papers with analogies to the game. A problem concerning "a maximal theorem with function-theoretic applications," he wrote in an important paper, could be "most easily grasped when stated in the language of cricket. . . . Suppose that a batsman plays, in a given season, a given 'stock' of innings." His words of praise for a great proof were not "straight from the Book," but "in the Hobbs class," after Surrey cricketer Jack Hobbs. When the Australian batsman Don Bradman eclipsed Hobbs's record, Hardy had to change the accolade: "Bradman is a whole class above any batsman who has ever lived: if Archimedes, Newton, and Gauss remain in the Hobbs class, I have to admit the possibility of a class above them, which I find difficult to imagine. They had better be moved from now on into the Bradman class." The barrage of references to cricket, said James Newman, "must not have been very helpful to the Hungarians," let alone to American mathematicians like himself.

Hardy was much more disciplined in his working style

than Erdős. He devoted only four hours each day, from 9:00 A.M. until 1:00 P.M., to mathematical research, leaving the afternoons open for cricket and tennis and the evenings for spirited discourse with the likes of Russell, Snow, philosopher G. E. Moore, logician Alfred North Whitehead, economist John Maynard Keynes, historian G. M. Trevelyan, novelist E. M. Forster, biographer Lytton Strachey, and publisher Leonard Woolf, among others. At Trinity College, Hardy had been a member of the Apostles, the famous student discussion group in which homosexuality—the "Higher Sodomy," as the Apostles called it—was not only accepted but treated as a more spiritual form of love. Homosexuality was not a membership requirement but so dominated the Apostles that "even the womanisers pretend to be sods, lest they shouldn't be thought respectable." Not known to have a boyfriend or a girlfriend, Hardy was apparently asexual, "a nonpracticing homosexual," as Littlewood put it.

Hardy was not only a child prodigy—at the age of two he was writing down numbers up to a million, and later, in church, was entertaining himself by finding the prime divisors of the hymn numbers—he was also a gifted athlete. "In his fifties he could usually beat the university second string at real tennis," said his friend C. P. Snow, "and in his sixties I saw him bring off startling shots in the cricket nets." His concerns were wide-ranging, as evidenced by six New Year's resolutions he sent in a postcard to a friend: "(1) prove the Riemann Hypothesis; (2) make 211 not out in the fourth innings of the last Test Match at the Oval; (3) find an argument for the nonexistence of God which shall convince the general public; (4) be the first man at the top of Mount Everest; (5) be proclaimed the first president of the U.S.S.R. of Great Britain and Germany; and (6) murder Mussolini."

Since the days of Newton, mathematics in Great Britain had taken a backseat to physics. In fact, pure mathematics was moribund there in 1877, the year Hardy was born. With his powerful analytical methods, and eye for beautiful patterns, Hardy succeeded in turning Britain into a world leader in pure mathematics for the first quarter of this century, an accomplishment akin to taking a cricket team from last place to first in a single season. Like Erdős, Hardy was a master at collaboration, although he worked with far fewer people but did so much more intensely. In 1911, Hardy began working with Littlewood, in a partnership that was unparalleled in the history of mathematics. Over a period of thirty-five years, they wrote one hundred joint papers, none of them "trivial."

When Hardy and Erdős met, Hardy was fifty-seven and his analytical powers were beginning to fade. Six years later he would write *A Mathematician's Apology*, the most famous literary work in mathematics, a depressing testament to creative powers that had slipped away. "It is a melancholy experience for a professional mathematician to find himself writing about mathematicians," the book begins. "The function of a mathematician is to do something, to prove new theorems, to add to mathematics, and not to talk about what he or other mathematicians have done." Make no mistake, Hardy warned, mathematics is a young man's game: "Galois died at twenty-one, Abel at twenty-seven ... Riemann at forty. ... I do not know an instance of a major mathematical advance initiated by a man past fifty." The twenty-one-year-old Erdős was too young then to know he would become the most celebrated counterexample to Hardy's conjecture.

Erdős asked Hardy what his most important contribution to mathematics was. "The discovery of Ramanujan,"

Hardy immediately responded. Besides his partnership with Littlewood, Hardy's only other intense collaboration was four years of work with the legendary Srinivasa Ramanujan Aiyangar, a poor, sickly, self-taught Indian phenom whose passage to England he arranged for in 1914, when Ramanujan was twenty-six. Erdős wanted Hardy to tell him about Ramanujan because their paths had crossed mathematically even though their lives had not. Erdős had first learned about Ramanujan's work in 1931, when he offered up his clever proof of Chebyshev's theorem that there is always a prime between a number and its double. A Hungarian colleague told him that Ramanujan had found a similar proof in 1919, so Erdős looked it up and was taken with the beauty of Ramanujan's approach.

Hardy told Erdős that he discovered Ramanujan when a letter postmarked "Madras, 16th January 1913" crossed his desk. Accompanied by pages of formulas that looked somewhat familiar to Hardy but were loaded with strange symbols, the letter began:

> Dear Sir,
>     I beg to introduce myself to you as a clerk in the Accounts Department of the Port Trust Office at Madras on a salary of only £20 per annum. I am now about 23 years of age. I have no University education but I have undergone the ordinary school course. After leaving school I have been employing the spare time at my disposal to work at Mathematics. . . . I am striking out a new path for myself. I have made a special investigation of divergent series in general and the results I get are termed by the local mathematicians as "startling."

Hardy wondered whether the correspondent was a crank but he read on.

Very recently I came across a tract by you styled *Orders of Infinity* in page 36 of which I find a statement that no definite expression has as yet been found for the number of prime numbers less than any given number. I have found an expression which very nearly approximates the real result, the error being negligible.

Ramanujan was claiming in effect that he had found a superior version of the legendary Prime Number Theorem. Hardy got a chuckle out of this presumption and put the letter aside. But the strange formulas haunted him throughout the day, and so later that evening, at High Table, he shared the letter with Littlewood. Crackpot or genius? Littlewood wondered. For two and a half hours the most successful team in all of mathematics pored over the formulas. Genius, they concluded.

Ramanujan, it turns out, had taught himself mathematics, borrowing a book from the library called *Synopsis of Elementary Results in Pure Mathematics*, written by an intellectual hack named George Shoebridge Carr, who eked out a living helping students cram for exams. The *Synopsis* listed some 6,000 formulas, none of them proved. Ramanujan took it upon himself to generate all the formulas anew, inventing his own notation to do this, performing calculations on a slate because he could hardly afford paper, and entering the final results in private notebooks he kept. Because the *Synopsis* didn't include proofs, Ramanujan didn't know the concept of a formal mathematical argument. He winged it, relying on intuition, but intuition he had in spades. Prime numbers, to be sure, were Ramanujan's friends as well as Erdős's.

In his reply, Hardy, the master of mathematical formality, pushed Ramanujan for proof of his fantastic conjectures. Proof meant everything to Hardy, who once told

Russell "that if he could find a proof that I [Russell] was going to die in five minutes he would of course be sorry to lose me, but this sorrow would be quite outweighed by pleasure in the proof. I entirely sympathised with him and was not at all offended." Ramanujan was delighted that Hardy responded; he had never heard from two other well-known English mathematicians to whom he had sent his results. But when it came to this thing called proof, Ramanujan did not know exactly what Hardy was asking for, and he had no interest in rehashing the logic behind assertions he *knew* were true. Life was too short, he felt, to go over the same ground twice. Little did he know how short. Six years after Hardy had persuaded Ramanujan to come to England, he died of tuberculosis in 1920 at the age of thirty-two, leaving a widow, no children, and three notebooks full of mathematical musings that experts are still deciphering today. To the excitement of number theorists around the world, a fourth "lost" notebook was discovered in 1976. The notebooks contain hundreds of insightful formulas, all offered without proof, as if they had been handed down by God.

While Hardy and Ramanujan's partnership lasted, the two men stood the world of pure mathematics on its head. It was East meets West, mysticism meets formality, and the combination was unstoppable. Ramanujan had flashes of insight, guided at night by visions of the goddess Namagiri, whose timely intervention had allowed Ramanujan to defy the Brahmin ban on overseas travel and meet Hardy in the first place. "An equation for me," Ramanujan said, "has no meaning unless it expresses a thought of God." Ramanujan had glimpsed the SF's Book, and Hardy turned the glimpses into proofs. "I owe more to him than to anyone else in the world with one exception," said Hardy, "and my association with him is the one romantic incident in my life."

Ramanujan, for his part, was not happy in England. He was lonely and homesick, craved Indian food, hated the weather, and was constantly sick. Even though he had discovered profound new mathematical truths, he was distressed that, for lack of a formal European mathematical education, he also kept rediscovering theorems that had already been proved. Early in 1918, Ramanujan threw himself onto the London subway tracks in front of an oncoming train. "What happened next," wrote his biographer, Robert Kanigel, "would be easy enough to read as a miracle. A guard spotted him . . . and pulled a switch, bringing the train screeching to a stop a few feet in front of him. Ramanujan was alive, though bloodied enough to leave his shins deeply scarred. He was arrested and hauled off to Scotland Yard." He didn't try to commit suicide again, but was never really himself—tuberculosis was taking over his body.

Hardy liked to rank mathematicians on a scale of 1 to 100, said Erdős. He gave himself 25, Littlewood 30, the great David Hilbert 80, and Ramanujan 100. "Although Hardy was modest in giving himself just 25," said Erdős, "the fact that he gave 100 to Ramanujan revealed the regard he had for Ramanujan's work."

Ramanujan, said Hardy, "could remember the idiosyncrasies of numbers in an almost uncanny way. It was Littlewood who said that every positive integer was one of Ramanujan's personal friends. I remember going to see him once in Putney. I had ridden in taxi-cab No. 1729, and remarked that to me the number seemed a rather dull one, and that I hoped it was not an unfavourable omen. 'No,' he reflected, 'it is a very interesting number; it is the smallest number expressible as the sum of two cubes in two different ways.'" Ramanujan had seen that

$$1729 = 12^3 + 1^3 = 10^3 + 9^3$$

Erdős liked hearing Hardy talk about Ramanujan. The two times Erdős would lecture in India, he'd have the fee donated to Ramanujan's widow, a woman he never met. And Hardy presumably liked hearing Erdős talk. He and his fellow Brits got a kick out of Erdős's Hungarian-sounding English, in which "pineapple upside down cake" became "pinnayopp-play oopshiday dovn tsockay."

Erdős shared Ramanujan's fondness for finding so-called asymptotic formulas, of which the Prime Number Theorem is a famous example. An *asymptotic formula* gives an estimate of the number of numbers below some arbitrarily large integer $n$ that have a certain desired property. The farther out you go in the integers—in other words, the larger you choose $n$—the more accurate the estimate becomes as a percentage of the true value. The Prime Number Theorem gives an estimate of the number of primes up to an integer $n$. As Gauss discovered, the estimate gets better as $n$ gets larger, to the point where if you're dealing with sufficiently high numbers, the estimate and actual value get "asymptotically" close—that is, they converge.

| *Integers below* n | Actual number of primes | Gauss's estimated number of primes | Percentage of error |
|---|---|---|---|
| 1,000 | 168 | 145 | 16.0 |
| 1,000,000 | 78,498 | 72,382 | 8.4 |
| 1,000,000,000 | 50,847,478 | 48,254,942 | 5.4 |

Ramanujan was mistaken in claiming in his letter to Hardy that he had improved on Gauss. But he did come up with asymptotic formulas for all sorts of other things that Gauss didn't even consider.

In his first year at Cambridge, Ramanujan worked on composite numbers, which are numbers that are not prime, numbers that are "composed" of the products of primes. Thus 6 is composite, because it is the product of 2 and 3. So is 15, being the product of 3 and 5. Ramanujan introduced the concept of a "highly composite number," which was as dissimilar to a prime as a number can be. A prime has only two divisors (the number 1 and itself), whereas a highly composite has the maximum number. A highly composite is characterized by having more divisors than any composite less than it has. Twelve is highly composite because it has six distinct divisors (1, 2, 3, 4, 6, and 12), which is more than any composite less than it has. So is 24, with its eight divisors (1, 2, 3, 4, 6, 8, 12, and 24) being greater in number than the divisors of any smaller composite. Ramanujan made a list of all highly composite numbers up to 6,746,328,388,800. The list began 2, 4, 6, 12, 24, 36, and included twenty-five highly composites between 2 and 50,000. The list was correct, except that he overlooked 29,331,862,500. Ramanujan found an impressive asymptotic formula that gives the number of highly composite numbers up to an arbitrarily large integer $n$. Thirty years later, in 1944, Erdős was able to improve on the formula.

Since the time of Euclid, mathematicians have known that every composite number can be uniquely expressed as the product of prime factors:

$$2^a \times 3^b \times 5^c \times 7^d \ldots$$

where the exponents $a$, $b$, $c$, $d$, and so on take on integral values. Ramanujan examined the highly composite numbers expressed as the products of primes. For instance,

$$6 = 2^1 \times 3^1$$
$$12 = 2^2 \times 3^1$$
$$24 = 2^3 \times 3^1$$

He noticed that in all three examples, the final exponent was 1. He realized that this was true of all highly composite numbers (with only two exceptions, 4 and 36, which can be expressed respectively as $2^2$ and $2^2 \times 3^2$). He noticed, too, that in all three of these prime factorizations, 2 is raised to an equal or higher power than 3 is. Ramanujan generalized this discovery, claiming that in the prime factorization of any highly composite number, the first exponent $a$ always equals or exceeds the second exponent $b$, which always equals or exceeds any third exponent $c$, and so on. This held for every case he examined. Witness:

$$332,640 = 2^5 \times 3^3 \times 5^1 \times 7^1 \times 11^1$$
$$43,243,200 = 2^6 \times 3^3 \times 5^2 \times 7^1 \times 11^1 \times 13^1$$
$$2,248,776,129,600 = 2^6 \times 3^3 \times 5^2 \times 7^2 \times 11^1 \times$$
$$13^1 \times 17^1 \times 19^1 \times 23^1$$

Ramanujan wrote up his explorations of highly composite numbers in a fifty-two-page paper that earned him his B.A. from Cambridge in March 1916, the university having waived the requirement that he take courses because he preferred to sit alone in his room, eating lentils and ghee while probing the secrets of primes. The simple questions he worked on were endless. "Hardy liked to say," said Erdős, "that every fool can ask questions about prime numbers that the wisest man cannot answer."

In 1917, Hardy and Ramanujan wrote a paper on so-called round numbers, which were composites that had an abnormally high number of prime divisors compared with other composites of the same magnitude. One measure of a number's roundness is to count the number of times each prime divisor appears in the number's prime factorization. By this measure, 1 million, whose prime factorization is $2^6 \times 5^6$, has a roundness of 12 (the sum of the exponents 6 and 6). Composite numbers between 991,991 and 1,000,010 have on average four prime divisors. So 1 million, with three times as many prime divisors, turns out to be very round:

| Number | Roundness | Prime Factorization |
|---|---|---|
| 999,991 | 3 | $17 \times 59 \times 997$ |
| 999,992 | 6 | $2^3 \times 7^2 \times 2,551$ |
| 999,993 | 2 | $3 \times 333,331$ |
| 999,994 | 3 | $2 \times 23 \times 21,739$ |
| 999,995 | 2 | $5 \times 199,999$ |
| 999,996 | 5 | $2^2 \times 3 \times 167 \times 499$ |
| 999,997 | 2 | $757 \times 1,321$ |
| 999,998 | 4 | $2 \times 31 \times 127^2$ |
| 999,999 | 7 | $3^3 \times 7 \times 11 \times 13 \times 37$ |
| 1,000,000 | 12 | $2^6 \times 5^6$ |
| 1,000,001 | 2 | $101 \times 9,901$ |
| 1,000,002 | 3 | $2 \times 3 \times 166,667$ |
| 1,000,004 | 5 | $2^2 \times 53^2 \times 89$ |
| 1,000,005 | 4 | $3 \times 5 \times 163 \times 409$ |
| 1,000,006 | 3 | $2 \times 7 \times 71,429$ |
| 1,000,007 | 2 | $29 \times 34,483$ |
| 1,000,008 | 8 | $2^3 \times 3^2 \times 17 \times 19 \times 43$ |
| 1,000,009 | 2 | $293 \times 3,413$ |
| 1,000,010 | 4 | $2 \times 5 \times 11 \times 9,091$ |

Hardy and Ramanujan came up with an asymptotic formula for roundness. Twenty-two years later, Erdős, working with Mark Kac, one of the founders of modern probability theory, would find a deep connection between a number's roundness and that workhorse of probability theory the bell-shaped curve, or normal distribution. It is testimony to the power and breadth of mathematics that such diverse subjects, prime number theory and probability theory, are actually linked.

Hardy and Ramanujan's most famous joint paper was in the theory of partitions, the ways of representing a given whole number $n$ as the sum of positive whole numbers. The number 5, for example, can be "partitioned" seven ways:

$$5 = 5$$
$$5 = 4 + 1$$
$$5 = 3 + 2$$
$$5 = 3 + 1 + 1$$
$$5 = 2 + 2 + 1$$
$$5 = 2 + 1 + 1 + 1$$
$$5 = 1 + 1 + 1 + 1 + 1$$

As $n$ grows, the number of partitions balloons. For $n = 10$, the partitions number 42. For $n = 50$, they number 204,226. For $n = 100$, they number 190,569,292. And for $n = 200$, a whopping 3,972,999,029,588. In 1918, in a forty-page paper on partition theory, Hardy and Ramanujan offered a surprisingly accurate asymptotic formula for the number of partitions of an integer $n$. In 1942, Erdős was able to show that Hardy and Ramanujan didn't need to use heavy machinery to deduce the first term of their formula, that the term could be found by "elementary" methods.

Elementary techniques are not necessarily simpler. In this context, elementary means that the proof of the formula relies on a restricted set of numbers, the so-called real numbers, which consist of the integers, the rational numbers, and the irrational numbers. The integers include all whole numbers: 0, the counting numbers, 1, 2, 3 . . . , and the negative counting numbers −1, −2, −3. . . . A *rational* number, from the word *ratio*, is just a fancy name for a fraction, like 2/3 or 3/5. When represented in decimal form, a rational number always terminates (like 1/4 represented as .25) or resolves itself into a pattern that repeats itself over and over (say, 1/3 as .3333333 . . . or 1/7 as .142857142857142857142857 . . . ). An *irrational* number, on the other hand, has a decimal notation that goes on forever without a repeating pattern of digits. The irrationals include $\sqrt{2}$ (or 1.4142135623 . . . ) and $\pi$ (or 3.1415926535 . . . ). What's not admitted are so-called imaginary numbers like $\sqrt{-1}$, which, when multiplied by itself, equals −1. The $\sqrt{-1}$, also known as *i*, is *imaginary* because it defies the conventional wisdom that the multiplication of two positives or two negatives is always a positive. Erdős didn't have philosophical objections to imaginary numbers; he just preferred to limit his toolkit to the more familiar integers, rationals, and irrationals. Erdős was the master of elementary methods. When he visited Hardy in 1934, Hardy believed that the Prime Number Theorem would never yield to elementary methods, and that should it ever yield, number theory textbooks would have to be thrown out and rewritten from scratch. Hardy would die in 1949, only months before Erdős and Selberg commandeered elementary methods to slay the Prime Number Theorem.

■

Erdős's four years in England were mathematically pleasant, "but I was very homesick," he recalled, "so I went back to Budapest three times a year, for Easter, Christmas, and the summer." On the trips home he saw his parents and got together with friends like Vázsonyi.

"His absence was hard on me mathematically," said Vázsonyi. "I had had no one to talk to. He was the only one in our circle who talked to me about the field of mathematics for which I had a natural affinity, the field of projective geometry"—the study of the properties of shapes that stay the same when the shapes are projected onto surfaces. "Math has fashions. People are interested in certain things. In the nineteenth century, volumes and volumes were published on elliptic functions." Then elliptic functions fell out of favor for decades, until interest surged a bit recently in connection with such endeavors as computing far-out digits of $\pi$. "Projective geometry was already a dead field by the time I came to it," said Vázsonyi. "I was blessed—or should I say cursed?—with talents in an area no one cared about anymore. Except Erdős, who indulged my interests."

In 1936, on one of Erdős's visits to Budapest, Vázsonyi "was doing research on a classical graph theorem, the Königsberg theorem of Euler, and managed to extend the theorem to infinite graphs. I had only the necessary but not the sufficient condition," he recalled. "I used to meet with Erdős practically daily and made the fatal mistake of telling Erdős on the phone about my discovery. I say fatal because he called me back in 20 minutes and told me the proof of sufficient condition. 'Damn it,' I thought, 'now I have to write a joint paper with him.' Little did I know the fame Erdős number one would bring me."

When Erdős left, though, a void set in. "I couldn't follow what Turán was working on," Vázsonyi bemoaned. "Then this more-pressing problem came up: how to save my life, how to escape from Hungary."

# e

---

# PROBLEMS WITH SAM AND JOE

There is a much quoted story about David Hilbert, who one day noticed that a certain student had stopped attending class. When told that the student had decided to drop mathematics to become a poet, Hilbert replied, "Good—he did not have enough imagination to become a mathematician."

—Robert Osserman

What men are poets who can speak of Jupiter if he were like a man, but if he is an immense spinning sphere of methane and ammonia must be silent?

—Richard Feynman

On March 13, 1938, Austria surrendered to Hitler, bringing the Third Reich to the western border of Hungary, a short one hundred miles from Budapest. "It was too dangerous for me to return to Hungary in the spring," said Erdős. "I did slip back in during the summer. But on September 3, I didn't like the news—the Czech crisis—so I went back to England that evening and was on my way to the United States three and a half weeks later." He went first to Princeton, New Jersey, for a low-paying fellowship at the Institute for Advanced Study. But after one year the Institute leadership only renewed his fellowship for six months because they found him uncouth and unconventional. The irony was that when Erdős looked back on his career in 1975, he fondly recalled his late 1930s stay at the Institute as his most productive time ever mathematically. World peace and tranquility, evidently, were not prerequisites for his mathematical genius. In 1939, the order of the world around him was crumbling, and the safety of his family and friends loomed large in his mind. On September 1, Hitler attacked Poland and World War II broke out. At the time Horthy refused to allow German troops to cross Hungary on their way to Poland, but his resolve to resist the Nazis was wearing down.

In the 1940s, Erdős collaborated with the set theorist Stanislaw Ulam, whom he had met in 1935 in Cambridge, England. Like Erdős, Ulam was a child prodigy, who had proved before the age of twenty a key result about the nature of infinite sets. "When Stan heard that Erdős's fellowship at the Institute wasn't renewed," recalled his widow, Françoise Ulam, "he tried to help him out by inviting him to give a talk in Madison, Wisconsin, where Stan was teaching." Erdős happily took up the offer.

Erdős in 1941 was "twenty-seven years old, homesick, unhappy, and," wrote Stan Ulam in his autobiography, "constantly worried about the fate of his mother who had remained in Hungary." Erdős could no longer correspond with her once Horthy made the fateful decision on June 22 to join the Nazis in attacking the Soviet Union and mail between the United States and Hungary was cut off. Erdős's visit to Madison "became the beginning of our long intense—albeit intermittent—friendship," recalled Ulam.

"Being hard-up financially—'poor,' as he used to say—he tended to extend his visits to the limits of welcome." On that visit and a subsequent one, "we did an enormous amount of work together," Ulam said, "our mathematical discussions being interrupted only by reading newspapers and listening to radio accounts of the war or political analyses." Erdős's appearance made an impression on Ulam: "Erdős is somewhat below medium height, an extremely nervous and agitated person . . . almost constantly jumping up and down or flapping his arms. His eyes indicated he was always thinking about mathematics, a process interrupted only by his rather pessimistic statements on world affairs, politics, or human affairs in general, which he viewed darkly. If some amusing thought occurred to him, he would jump up, flap his hands, and sit down again."

Ulam left Madison in 1943 to join the physicists at Los Alamos, New Mexico, who were secretly building atomic weapons. The application of mathematics was new to Ulam. He told Otto Frisch, a physicist at Los Alamos, "that he was a pure mathematician who used to work entirely with abstract symbols, but had now sunk so low that his latest report had contained actual numbers, indeed numbers with decimal points; that (he pretended) was the ultimate disgrace! In fact he had an uncanny skill of using the most abstruse and abstract techniques of mathematics for predicting the behaviour of an atomic bomb." Ulam invited Erdős to join the war effort and suggested he write to his fellow countryman Edward Teller, who had immigrated to the United States in 1935. Erdős wrote to "Professor Edwards" but was disqualified when he made a big point of saying that he might return to Budapest after the war. He loved tweaking the authorities. To Peter Lax, another Hungarian émigré at Los Alamos, he fired off a short postcard:

"Dear Peter, my spies tell me that Sam is building an atomic bomb. Tell me, is that true?"

According to Richard Bellman, a mathematician at Los Alamos, Erdős came to nearby Santa Fe to plead his case for joining them. "He was always strongly anti-Fascist and wanted to work at Los Alamos where his great talents would have been very useful. Unfortunately, he refused to sign a paper saying that he would not talk about the A-bomb after the war. Considering all the foreign scientists at Los Alamos such a paper was a farce. We took him to dinner at the La Fonda, Betty Jo, I, Peter Lax and John Kemeny, another young soldier of Hungarian background. Erdős spoke in Hungarian about the relatives of Peter and John back in Hungary. Then, in loud English he asked, 'How is work going on the A-bomb?' "

Even if he had not disclosed that he was homesick for Hungary and had promised a vow of silence after the war, Erdős probably would not have been invited to Los Alamos. The FBI already had a file on him because of a false arrest two summers before. Erdős had been picked up on Long Island for loitering suspiciously near a military radio transmitter. His run-in with the G-men made the New York tabloids. 3 ALIENS NABBED AT SHORT-WAVE STATION screamed a *Daily News* headline on August 15, 1941. The *Post* story, under the more restrained headline FBI DEFEATS SPY SCARE: 3 ALIENS JUST STUDENTS, struck a bemused tone:

Riverhead, L.I., Aug. 15—It's all right about those three highly suspicious aliens who were reported to be snooping around the Mackay Radio station in Southampton yesterday. The FBI says so. The FBI ought to know. It questioned them half the night.

It seems the employees of the station telephoned

police to say that a Japanese and two others (they turned out to be a Hungarian and an Englishman) were making sketches of a 200-foot tower. The three went away in a car when they were questioned, said the radiomen.

## FATHER OF ONE A CENSOR

The three were picked up later in Easthampton 14 miles away. They said they were Shizuo Kakutani, 29, Paul Erdős, 28, and Arthur Harold Stone, 22.

Kakutani said he was a mathematics student in the Princeton graduate school. Erdős said he had a fellowship in mathematics. . . . Stone said his father was an English postal censor and he also was studying at Princeton.

## THEIR FILMS ALL BLANK

All said they were on their way to a scholastic conference in Chicago and decided to stop off and look at the Long Island seacoast. All denied any but casual interest in the short-wave transmitter of the Mackay Co.

That's about all. They are students. They had little interest in the transmitter. And the slightly suspicious camera found in their car was found to contain 10 exposed films—all blank.

"It was a harmless situation," recalled Erdős. "I certainly can't blame the United States. That was long before the McCarthy era, and they all reacted reasonably," except for an overzealous guard, "who wasn't terribly clever." Stone was taking Erdős and Kakutani on a tour of Long Island. They overlooked a NO TRESPASSING sign and drove up to what looked like a radio transmitter but actually may have been a secret radar facility.

The three men got out of the car to stretch and look at the ocean. They took a few photos of each other. As they were getting back into the car, the guard told them to leave, which they promptly did. He then got spooked and reported that three Japanese men had taken notes, snapped photos, and "left with sudden suspiciousness." Of course, he had ordered them to leave, and "there was only one Japanese," said Erdős, "we didn't take any notes, and the photographs which we took were people." They had been talking mathematics the whole time, and the guard may have been worried because he didn't understand a word.

"It was quite harmless," said Erdős, "but they raised a nine-state alarm, and looked for us all over. We didn't suspect anything, so we went to the end of Long Island.... At lunch we were arrested. At that time, the two detectives probably knew that the whole thing is nonsense, because they listened to our conversation and they realized that this must be all a mistake. But by that time it was too late because already the FBI was called." The FBI wanted to know how they could have missed the NO TRESPASSING sign. "I was thinking," Erdős told them. "What about?" they demanded. "Mathematics," he said. "By the evening, they cleared up everything," Erdős recalled, "and we were released.... You can find that in many newspapers of the time. Only the New York *Daily News* was unpleasant. They claimed that we were seen rowing near a naval base, which was a complete invention."

In 1943, Erdős took a part-time appointment at Purdue University and, as Ulam recalled, "was no longer entirely penniless—'even out of debts,' as he called it." At Purdue, Erdős had another minor run-in with the authorities. He liked to take long walks at all hours of the day and night, and the police once stopped him on a midnight stroll. Their suspicions were further aroused because he had no ID, no driver's license, nothing with his name on it. "What are you doing?" they asked. "I'm thinking," he said. "About what?" they pressed. "About mathematics," he said. Puzzled, they let him go.

At the Purdue math department, Erdős was a star. "While he meant much to our group because of his mathematical genius," recalled Michael Golomb, "he also contributed greatly to our social life. We all felt socially deprived, confined to a small town in a midwestern agricultural state, far from cosmopolitan centers. Now we had among us this man, with his past from European cultural centers, his intimate acquaintance with internationally known mathematicians, but also his wide interest in scientific culture and world politics." The Purdue mathematicians gathered once a week with colleagues from other departments for an informal talk followed by a lively, free-for-all discussion. One time the speaker did not show up, "and no other program had been prepared. Erdős himself offered to speak. Extemporaneously, without the aid of notes, he gave a fascinating report about some recent research on the color vision of bees. We were enchanted and surprised by this performance; we had not expected that Paul had any interest in such matters."

The Purdue mathematicians were not the only ones who discovered the depths of Erdős's nonmathematical knowledge. Although numbers were his great love, and he had a short attention span for everything else, Erdős was by no means an idiot savant. In his youth, before he started putting in nineteen-hour days doing math, he read about history, science, and politics, and what he read stuck with him. Years later he would dazzle people with arcane historical and scientific knowledge. János Pach remembers when Erdős was introduced to Lajos Elekes, at the guesthouse of the Hungarian Academy of Sciences in Mátraháza.

"What's your profession?" Erdős began, a standard opening question whenever he met someone new (unless that person was accompanied by an infant, in which case his first questions inevitably were, "How old is the epsilon? Is it a boss or a slave?"). Many people would be short of words upon learning that Elekes was a historian who was writing a book about János Hunyadi, a fifteenth-century Hungarian general. Erdős immediately pressed Elekes on "the causes of the catastrophic defeat of the Hungarian army by the Turks in the battle of Varna, in 1444."

■

For the first two and half years of World War II, Hungarian Jews did fairly well, despite random acts of anti-Semitic violence and various legal and economic restrictions that Horthy imposed in order to appease Hitler. But on March 19, 1944, the Nazis invaded Hungary and with lightning speed systematically liquidated Hungarian Jews in what Winston Churchill called "probably the greatest and most horrible crime ever committed in the history of the world." By July 7 more than 437,000 Jews, 50,000 of them from Budapest, had been deported to Auschwitz. That winter, 20,000 of the 160,000 Jews remaining in the Budapest ghetto, where Erdős's mother had been forced to move, were either murdered or brought down by starvation, cold, and disease. On February 14, 1945, Soviet troops liberated Budapest. In August, Erdős, who had not heard a word from his parents since 1941, received a telegram. His mother was alive, and his cousin Magda Fredro had survived Auschwitz, but the rest of the news wasn't good. "The Nazis ended up murdering four of my mother's five brothers and sisters," said Erdős, "and my father died of a heart attack in 1942."

The Soviet authorities, who would control Hungary until 1988, lost no time in replacing the Nazi horror with a terror of their own. "Under orders from Stalin to fill quotas of prisoners to be sent to the GULAG, the majority never to return, the liberators pick men in the streets at random for the *malenkiĭ robot* ['small labor']," wrote Lázló Babai. "Fresh from the ordeal of 'labor companies' and *Nyilas* [Hungarian Nazi gang] terror, Turán is stopped by a Soviet patrol in Budapest. The officer demands his papers. Having escaped from a *Nyilas* round-up a few days earlier, Turán has no proper ID. He does carry, however, a reprint of his 1935 paper with Erdős, published in the *Bulletin of the Institute of Mathematics and Mechanics of Tomsk*. . . . The officer is impressed by the prewar publication in a Soviet journal and lets Turán go. 'An unexpected application of number theory,' Turán reports to Erdős." (Twenty-eight years later, in Los Angeles, number theory would similarly save Erdős from U.S. authorities. He was arrested for jaywalking, and with no money or ID on him, the police threatened to haul him into jail. He showed them his weighty volume of collected papers, *The Art of Counting*, with his beaming picture on the frontispiece. They shrugged and accepted it as photo ID, and a colleague at UCLA, Ramsey theorist Bruce Rothschild, paid the fine.)

Ulam did not make it through the war years unscathed. In the winter of 1945 he was rushed to a hospital in Los Angeles, suffering a violent headache, numbness in his chest, and slurred speech. "I remembered suddenly Plato's description of Socrates after he was given the hemlock in prison," Ulam later wrote. "The jailor made him walk and told him that when the feeling of numbness starting in the legs reached his head he would die." The doctors found that Ulam's brain was severely inflamed, "bright pink instead of the usual gray." They gave him antibiotics and drilled a hole in his skull to relieve the pressure. When the swelling subsided and Ulam awoke after a few days of postoperative coma, his surgeon tested his mental faculties by asking him the sum of 8 and 13. "The fact that he asked such a question embarrassed me so much that I just shook my head," recalled Ulam. "Then he asked me what the square root of twenty was, and I replied: about 4.4. He kept silent, then I asked, 'Isn't it?' I remember Dr. Rainey laughing, visibly relieved, and saying, 'I don't know.'" The authorities at Los Alamos were also concerned about Ulam's mental state; they worried that while unconscious he might have given away atomic secrets. They wondered, too, whether exposure to atomic radiation could have caused the encephalitis, but concluded that he had never been sufficiently close to radioactive material.

After a few weeks in the hospital, he was strong enough to leave. As his wife Françoise was leading him out of the hospital, he was greeted by a perky and happy Erdős: "Stan, I am so glad to see you are alive. I thought...I would have to write your obituary and our joint papers. You are going home? Good, I can go with you." Françoise was alarmed by this prospect. "I was afraid he'd tire Stan out," she recalled fifty years later. "For the whole car ride home he wouldn't let Stan rest but engaged him in mathematical conversation. And as soon as we got home, he challenged Stan to chess. At first I was horrified, but then I realized that there was method to Erdős's seeming madness: with these mental challenges, he was helping Stan recover his mathematical self-confidence." Stan was nervous about the chess game, fearing that he had forgotten the rules. And when he won the first game, Ulam worried that Erdős might have thrown it on purpose. Erdős wanted a rematch. "I agreed," Stan wrote in his autobiography, "although I felt tired, and won again. Whereupon it was Erdős who said, 'Let us stop, I am tired.' I realized from the way he said it that he had played in earnest." Erdős stayed at their house for two weeks, bombarding Ulam with mathematical chestnuts. "The stay was hard on me," said Françoise. "I had to care for both of them—but Stan loved his presence and I am forever grateful to Erdős for helping Stan regain his mathematical equilibrium."

Four decades later, Erdős did have to write Ulam's obituary, his death in 1984 ending a mathematical collaboration that had lasted fifty years. "Ulam was always afraid of getting old and was rather proud that he could play good tennis even when he was over 70," Erdős wrote. "He was really fortunate to have avoided the greatest evils, old age and stupidity, and he died suddenly of heart failure without fear or pain while he could still prove and conjecture. . . . In the 1001 nights, the king was greeted by 'O King may you live forever.' A mathematician and scientist can be greeted by the more realistic 'O Mathematician, may your theorems live forever.' I wish and expect this fate for Stan."

Elementary number theory, a subject dear to Erdős, was one of Ulam's earliest mathematical interests. In high school Ulam went on the hunt for an odd perfect number (of course, he didn't find one, but neither did anyone else). As a "dotigy"—Erdős and Ulam's term for the antipathy of prodigy—Ulam returned to elementary number theory somewhat by accident. A year before he died, in 1983, Ulam was stuck listening to "a long and very boring paper" at a scientific conference. He passed the time doodling and found himself scribbling consecutive integers, starting with 1, in a kind of counterclockwise spiral:

```
100  99  98  97  96  95  94  93  92  91

 65  64  63  62  61  60  59  58  57  90

 66  37  36  35  34  33  32  31  56  89

 67  38  17  16  15  14  13  30  55  88

 68  39  18   5   4   3  12  29  54  87

 69  40  19   6   1   2  11  28  53  86

 70  41  20   7   8   9  10  27  52  85

 71  42  21  22  23  24  25  26  51  84

 72  43  44  45  46  47  48  49  50  83

 73  74  75  76  77  78  79  80  81  82
```

He was surprised to see that the primes (shown in bold) tended to fall on diagonal lines.

Inspired by this chance discovery, Ulam wrote out other square spirals of consecutive integers. The Maniac II mainframe at Los Alamos had in its memory the first 90 million prime numbers, and the weapons lab had one of the first computer graphics facilities. Ulam and two Los Alamos colleagues, Mark Wells and Myron Stein, programmed Maniac II to plot a square-spiral diagram of all the integers up to 10 million. The primes continued their uncanny preference for diagonals.

Since Euclid, the brightest minds in mathematics have tried without success to find patterns to the primes and formulas for generating them. Could Ulam, with his spiral scribbles, be onto something? "Ulam's doodles in the twilight zone of mathematics are not to be taken lightly," Martin Gardner wrote in 1964. After all, "it was he who made the suggestion that led him and Edward Teller to think of the 'idea' that made possible the first thermonuclear device."

In one smaller plot, Ulam put 17 in the center and added all the integers from 17 to 272.

| | | | | | | | | | | | | | | | |
|---|---|---|---|---|---|---|---|---|---|---|---|---|---|---|---|
| 272 | **271** | 270 | **269** | 268 | 267 | 266 | 265 | 264 | **263** | 262 | 261 | 260 | 259 | 258 | **257** |
| 213 | 212 | **211** | 210 | 209 | 208 | 207 | 206 | 205 | 204 | 203 | 202 | 201 | 200 | **199** | 256 |
| 214 | 161 | 160 | 159 | 158 | **157** | 156 | 155 | 154 | 153 | 152 | **151** | 150 | **149** | 198 | 255 |
| 215 | 162 | 117 | 116 | 115 | 114 | **113** | 112 | 111 | 110 | **109** | 108 | **107** | 148 | **197** | 254 |
| 216 | **163** | 118 | 81 | 80 | **79** | 78 | 77 | 76 | 75 | 74 | **73** | 106 | 147 | 196 | 253 |
| 217 | 164 | 119 | 82 | **53** | 52 | 51 | 50 | 49 | 48 | **47** | 72 | 105 | 146 | 195 | 252 |
| 218 | 165 | 120 | **83** | 54 | 33 | 32 | **31** | 30 | **29** | 46 | **71** | 104 | 145 | 194 | **251** |
| 219 | 166 | 121 | 84 | 55 | 34 | 21 | 20 | **19** | 28 | 45 | 70 | **103** | 144 | **193** | 250 |
| 220 | **167** | 122 | 85 | 56 | 35 | 22 | **17** | 18 | 27 | 44 | 69 | 102 | 143 | 192 | 249 |
| 221 | 168 | 123 | 86 | 57 | 36 | **23** | 24 | 25 | 26 | **43** | 68 | **101** | 142 | **191** | 248 |
| 222 | 169 | 124 | 87 | 58 | **37** | 38 | 39 | 40 | **41** | 42 | **67** | 100 | 141 | 190 | 247 |
| **223** | 170 | 125 | 88 | **59** | 60 | **61** | 62 | 63 | 64 | 65 | 66 | 99 | 140 | 189 | 246 |
| 224 | 171 | 126 | **89** | 90 | 91 | 92 | 93 | 94 | 95 | 96 | **97** | 98 | **139** | 188 | 245 |
| 225 | 172 | **127** | 128 | 129 | 130 | **131** | 132 | 133 | 134 | 135 | 136 | **137** | 138 | 187 | 244 |
| 226 | **173** | 174 | 175 | 176 | 177 | 178 | **179** | 180 | **181** | 182 | 183 | 184 | 185 | 186 | 243 |
| **227** | 228 | **229** | 230 | 231 | 232 | **233** | 234 | 235 | 236 | 237 | 238 | **239** | 240 | **241** | 242 |

Again, the primes fell diagonally in line. Indeed, the principal diagonal running from the lower left to the upper right was entirely prime: 227, 173, 127, 89, 59, 37, 23, 17, 19, 29, 47, 73, 107, 149, 199, and 257. To prime stalkers these numbers were familiar. In the eighteenth century Euler had advanced the formula $n^2 + n + 17$, which, for successive values of $n$, yielded primes for $n = 0$ through $n = 15$. These sixteen primes, in fact, were the ones that showed up on the principal diagonal of Ulam's plot.

| $n$ | $n^2 + \ n + 17 =$ | prime |
|---|---|---|
| 0 | $0^2 + \ 0 + 17 =$ | 17 |
| 1 | $1^2 + \ 1 + 17 =$ | 19 |
| 2 | $2^2 + \ 2 + 17 =$ | 23 |
| 3 | $3^2 + \ 3 + 17 =$ | 29 |
| 4 | $4^2 + \ 4 + 17 =$ | 37 |
| 5 | $5^2 + \ 5 + 17 =$ | 47 |
| 6 | $6^2 + \ 6 + 17 =$ | 59 |
| 7 | $7^2 + \ 7 + 17 =$ | 73 |
| 8 | $8^2 + \ 8 + 17 =$ | 89 |
| 9 | $9^2 + \ 9 + 17 =$ | 107 |
| 10 | $10^2 + 10 + 17 =$ | 127 |
| 11 | $11^2 + 11 + 17 =$ | 149 |
| 12 | $12^2 + 12 + 17 =$ | 173 |
| 13 | $13^2 + 13 + 17 =$ | 199 |
| 14 | $14^2 + 14 + 17 =$ | 227 |
| 15 | $15^2 + 15 + 17 =$ | 257 |

Put $n = 16$ into the formula $n^2 + n + 17$, however, and you get a composite number, 289, which is $17^2$. Euler came up with another formula $n^2 + n + 41$, which, for successive values of $n$ from 0 to 39, generates only primes. But it fails for $n = 40$, yielding 1,681, or 41 squared.

Maniac II showed that Euler's formula $n^2 + n + 41$ was surprisingly good, generating primes under 10 million a respectable 47.5 percent of the time. Ulam found other formulas whose hit rates were almost as good as Euler's. But to the dismay of prime fans like Erdős, Ulam's doodles ultimately didn't amount to much. Despite the seeming bias of the primes toward diagonals, number theorists proved that no formula like Euler's can generate only primes. Ulam gave up this flight of fancy and in the last year of his life resumed his role as an elder statesman of the nuclear age, lecturing on the relation of science to morality. He was once asked what the world would be like if his work at Los Alamos had proved the impossibility of building an atomic bomb. The world, he said, would obviously be a safer place to live, "without the risk of suicidal war and total annihilation. Unfortunately, proofs of impossibility are almost nonexistent in physics. In mathematics, on the contrary, they provide some of the most beautiful examples of pure logic."

■

Erdős saw it as his personal mission to help colleagues maintain their mathematical edge. When they fell ill, as Ulam did, he challenged their minds back to health. Not all recoveries were as successful as Ulam's. Some of his ill colleagues regained their mathematical ability but, tragically, not their confidence. Jon Folkman was a brilliant young mathematician who worked at Rand, the think tank in Santa Monica. He was diagnosed with a brain tumor in the late 1960s. Evidently it had grown to such a large size that the prognosis for surgery wasn't good. His doctors felt that there was little chance they could remove it; or, if they could, he might become a vegetable. But against the odds the operation was successful. "Erdős and I visited him in the hospital afterwards," said Graham. "No sooner had we entered the hospital room when Erdős started challenging Jon with math problems. He had just come out of brain surgery and he was able to solve the problems! Sure, he was answering a little slowly but he was answering. After he went home, though, his personality changed from before the procedure. He was a little moodier and he believed he was losing his skills. The evidence, though, was that his mathematical ability was better than ever. To test himself, he'd look through all the open problems in conference proceedings and solve them sequentially one by one. That's incredible! Jon, like Gauss, had such high standards he never published some of his very good work. One day he bought a gun and shot himself. He was thirty-one. It was all very sad. His boss at Rand, D. Ray Fulkerson, blamed himself for not recognizing the depths of his troubles and doing something. 'Jon's suicide is often on my mind,' Fulkerson said. Later Fulkerson killed himself too."

Even great mathematicians suffer bouts of insecurity, fearful that their skills are slipping away or their proofs are not as sound or significant as they had thought. Bertrand Russell once shared "a horrible dream" with G. H. Hardy. Russell dreamt he was on "the top floor of the University Library, about A.D. 2100," said Hardy. "A library assistant was going around the shelves carrying an enormous bucket, taking down book after book, glancing at them, restoring them to the shelves or dumping them into the bucket. At last he came to three large volumes which Russell could recognize as the last surviving copy of [his magnum opus] *Principia Mathematica*. He took down one of the volumes, turned over a few pages, seemed puzzled for a moment by the curious symbolism, closed the volume, balanced it in his hand and hesitated. . . ."

The Austrian logician Kurt Gödel was one of the mathematical giants who lost his confidence, and Erdős tried to help him get it back. Erdős met Gödel at the Institute for

Advanced Study, the logician's principal home from 1933 until 1976. "Gödel I talked with a great deal," said Erdős. "He was certainly a remarkable intellect. He understood everything, even what he didn't work [on]. It is strange how little he published. He could certainly have done more things. I always argued with him" because his interests drifted toward metaphysics. "We studied Leibniz a great deal and I told him, 'You became a mathematician so that people should study you, not that you should study Leibniz.'"

In 1931, back in Vienna, the twenty-five-year-old Gödel stunned the scientific community by tearing asunder the very foundations of mathematics. He had managed to prove that any formal mathematical system robust enough to include the laws of arithmetic would be unable to prove its own consistency. His innocently titled paper, "On Formally Undecidable Propositions of *Principia Mathematica* and Related Systems," skewered Russell's work on the foundations of mathematics (so that the librarian of the twenty-second century could hold on to *Principia Mathematica* only as a historical curiosity, not as consistent mathematical truth). Russell had already delivered the death blow to another logician, Friedrich Ludwig Gottlob Frege, so perhaps it was only divine retribution that Russell's work collapsed too.

Work on the foundations of mathematics was all the rage at the turn of the century. Mathematical logicians like Frege and Russell were trying to build up all of mathematics in a completely rigorous way. The idea was to take nothing for granted, to prove everything from first principles, to deduce all of mathematics from as few self-evident axioms as possible. In building up elementary arithmetic, for example, they started with axioms like the so-called commutative law of addition, that addition does not depend on the order of the numbers summed. In other words, $a + b = b + a$ (as opposed to subtraction, which is not commutative, because $a - b$ doesn't equal $b - a$). That they even bothered to state such self-evident propositions was a testament to their rigorous approach. But they went further, spelling out the definitions of individual numbers like the number 4 rather than taking such definitions for granted. In this formal way of thinking, numbers were defined in terms of sets. What, for example, is the number 4? Look around you, said Frege, and you'll see fours everywhere. There's the set of suits in a deck of cards and the set of legs on a chair. Take all the sets with four members and put them together in one big set—this set of sets constitutes the concept of "fourness."

All this painstaking work, which might seem better directed at finding new mathematical truths, started because of a crisis in geometry that mathematicians feared might spread to arithmetic. From a handful of seemingly self-evident truths—that two points uniquely define a straight line; that all right angles are equal; that a straight line extends indefinitely in both directions—Euclid, in five books of his thirteen-volume *Elements*, proved hundreds of geometric theorems about points and lines in a plane; for instance, that the sum of the angles of any triangle is 180 degrees. Euclid had done much more than put geometry on a rigorous footing. He offered the solace of certainty for souls troubled by uncertainty. For centuries his geometry— like the work of Newton, Darwin, Freud, and Einstein— took on a life of its own in the popular consciousness of Western culture. "Originally viewed as both a tool and a model for research in mathematics and other sciences, *The Elements* gradually evolved into a basic component of a standard education—a piece of intellectual equipment that every young student was expected to wrestle with and internalize," observed the Stanford mathematician Robert Osserman in *Poetry of the Universe*. "In a world full of irrational beliefs and shaky speculations, the statements found in *The Elements* were proven true beyond a shadow of a doubt. . . . The astonishing fact is that after two thousand years, nobody has ever found an actual 'mistake' in *The Elements*—that is to say, a statement that did not follow logically from the given assumptions."

Euclidean geometry reigned for more than two millennia, but in the backrooms of the palace there were always stirrings of discontent. One of Euclid's self-evident truths, the so-called parallel postulate, didn't seem entirely self-evident. The *parallel postulate* was the idea that given a line in a plane, and a point not on that line, you can draw through that point exactly one line parallel to the initial line. Sounds reasonable enough; but a few cranky empiricists, while not claiming that the postulate was false, questioned how one could be so sure that the lines didn't intersect way out in space. The French mathematician Jean Le Rond d'Alembart, in a flight of hyperbole, called this "the scandal of geometry." The situation actually became closer to a scandal once mathematicians started replacing the parallel postulate with a contradictory axiom.

In 1829, Nicolai Ivanovich Lobachevsky (immortalized in Tom Lehrer's ballad as a plagiarist, though he was no such thing) offered the intuitively strange alternative that given a point and a line distinct from the point, at least two distinct parallels could be drawn through the point. And in 1854, Georg Friedrich Bernhard Riemann proposed another contradictory alternative: that there are no parallel lines whatsoever, that all lines eventually meet up at infinity! Though defying common sense, each of these alternative geometries, Lobachevskian and Riemannian, seemed as internally consistent as Euclidean geometry. In other words, none could be thrown out on the grounds of self-contradiction. To be sure, the consequences of these new geometries flatly contradicted theorems of Euclidean geometry. For instance, in Riemannian geometry, the sum of the angles of a triangle is more than 180 degrees. In fact, the sum varies with the size of the triangle, and comes arbitrarily close to 180 degrees as the triangle gets smaller.

"What nonsense!" people thought. "Get out a compass and measure the angles of triangles in the real world and you'll see that they add up to 180 degrees." "Don't be so cocky," the Riemannians responded. "In the real world, the triangles you've measured are all small—remember, Earth is just a cosmic speck in the infinite expanse of the universe. That's why the angles of all your triangles *seem* to sum to 180 degrees. And even in your small world, it would be small-minded to conclude, given the inevitable imprecision of measurement, that the sum is precisely 180 degrees and not, say, 179.99997 degrees." It was one thing to offer up an alternate to the parallel postulate as an academic exercise; it was quite another to claim that tried-and-true Euclidean geometry was not actually the geometry of nature. Scandalous indeed, non-Euclidean geometry was the talk of café society throughout Europe.

Ivan, the skeptical sibling in Fyodor Dostoyevsky's *The Brothers Karamazov* (1880), clung to Euclid's camp:

If God exists and if he really did create the earth then, as common knowledge tells us, he created it according to Euclidean geometry, while he created the human mind with an awareness of only three spatial dimensions. Even so, there have been and still are even today geometers and philosophers of the most remarkable kind who doubt that the entire universe or, even more broadly, the entirety of being was created solely according to Euclidean geometry, and who even make so bold as to dream that the two parallel lines which according to Euclid can on no account converge upon earth may yet do so somewhere in infinity. And so, my lad, I've decided that if I can't even understand that, then how am I to understand about God? ... Even if the parallel lines converge and I actually witness it, I shall witness it and say they have converged, but all the same I shall not accept it.

With geometry in such disarray, logicians like Frege were working to shore up arithmetic. In 1902, Frege had the satisfied disposition of a man who has just completed his major work. The second volume of his book, *The Foundations of Arithmetic*, was at the printer, and the advance word was good because the first volume had made a splash in mathematical circles. Satisfaction gave way to despair, however, when he learned from Russell of an unavoidable paradox in the concept of a set of sets, the very concept that lay at the heart of his program. "A scientist can hardly meet with anything more undesirable than to have the foundations give way just as the work is finished," Frege said later. "In this position I was put by a letter from Mr. Bertrand Russell, as the work was nearly through the press."

The paradox Russell found had "an affinity with the ancient Greek contradiction about Epimenides the Cretan, who said that all Cretans are liars": if Epimenides was telling the truth, he was lying, and if he was lying, he was telling the truth. "A contradiction essentially similar to that of Epimenides," Russell wrote in his autobiography, "can be created by giving a person a piece of paper on which is written: 'The statement on the other side of this paper is false.' The person turns the paper over, and finds on the other side: 'The statement on the other side of this paper is false.' It seemed unworthy of a grown man to spend time on such trivialities, but what was I to do?"

Thinking about the paradoxical Cretan led Russell to the idea

> that a class sometimes is, and sometimes is not, a member of itself. The class of teaspoons, for example, is not another teaspoon, but the class of things that are not teaspoons is one of the things that are not teaspoons. There seemed to be instances which are not negative: for example, the class of all classes is a class. . . . [This] led me to consider the classes that are not members of themselves; and these, it seemed, must form a class. I asked myself whether this class is a member of itself or not. If it is a member of itself, it must possess the defining property of the class, which is to be not a member of itself. If it is not a member of itself, it must not possess the defining property of the class, and therefore must be a member of itself. Thus each alternative leads to its opposite and there is a contradiction.

A few years later, Russell came up with a popularized version of his paradox. Imagine the Barber of Seville who shaves every man who does not shave himself. Does the Barber of Seville shave himself? If he does, he doesn't, and if he doesn't, he does. Try as Frege did, he could not circumvent Russell's cunning conundrum about the class of all classes.

David Hilbert, the leading mathematician of the time, was the cheerleader for rebuilding the foundations of mathematics to purge it once and for all of nettlesome paradoxes. "What we have experienced," Hilbert said, "with the paradoxes of set theory . . . never will happen again." Hilbert's words were taken as gospel. "Every mathematical problem can be solved," he said. "We are all convinced of that. After all, one of the things that attracts us most when we apply ourselves to a mathematical problem is precisely that within us we always hear the call: here is the problem, search for the solution, you can find it by pure thought, for in mathematics there is no *ignorabimus.*"

Russell and Alfred North Whitehead responded to Hilbert's call. Like Frege before them, they tried to build up all of mathematics from first principles in three impenetrable volumes of *Principia Mathematica.* The first volume was published in 1910. The project went along swimmingly for two decades, until young Gödel derailed it.

Gödel demonstrated that no complex mathematical system was complete. In other words, no matter what axioms are chosen, meaningful mathematical statements can be made whose truth or falseness can never be demonstrated within the system. It was now possible that some of Erdős's prized problems and the open conjectures of other mathematicians were immune to proof. Gödel's second discovery was even more devastating. He demonstrated that it was impossible to prove that any given complex mathematical system was consistent. In other words, you can never be sure that the set of axioms won't lead to a contradiction. On the Richter scale of mathematical discoveries, Gödel's was a 10. That mathematics was incomplete and possibly inconsistent was a body blow to those who saw mathematics as the most logical of logical systems, and few in the field didn't see it that way. In the wake of Gödel, most card-carrying mathematicians still clung to the belief that mathematics was in fact free of contradictions, even though they now knew they could never prove this. As André Weil, number theorist extraordinaire, put it: "God exists since mathematics is consistent, and the Devil exists since we cannot prove it."

Erdős was in Weil's camp. He didn't believe that contradictions would show up. As for completeness, there was an inexhaustible supply of great problems out there, and they were yielding all the time to hard thinking, so what if a few could never be solved in theory? Erdős was having too good a time solving problems to worry about the philosophical underpinnings of his enterprise.

Russell, on the other hand, was crushed:

> I wanted certainty in the kind of way in which people want religious faith. I thought that certainty was more likely to be found in mathematics than elsewhere. But I discovered that many mathematical demonstrations, which my teachers wanted me to accept, were full of fallacies.... I was continually reminded of the fable about the elephant and the tortoise. Having constructed an elephant on which the mathematical world could rest, I found the elephant tottering, and proceeded to construct a tortoise to keep the elephant from falling. But the tortoise was no more secure than the elephant, and after some twenty years of very arduous toil, I came to the conclusion that there was nothing more that *I* could do in the way of making mathematical knowledge indubitable.

Genius though he was, Gödel was not a poster boy for mathematical sanity. Obsessed with ghosts and demons and an imagined heart ailment, he checked himself in and out of sanitariums many times in his adult life for treatment of depression and acute anxiety. He was always a finicky eater, but as he got older he ate less and less, refusing to take food from anyone but his wife Adele, fearing that other people were secretly trying to poison him. At sixty-four he weighed only eighty-six pounds. In the middle of 1977, when Adele was hospitalized for major surgery, he stopped eating altogether, and by the following January starved himself to death at the age of seventy-one. In his dying days he had serious doubts that his life's work amounted to anything more than the discovery of another silly paradox à la Barber of Seville. He was plagued by Russell's nightmare of future librarians trashing his work.

Gödel was not the only one at the Institute who had overturned the foundations of his subject. His fellow resident and friend Albert Einstein had done that in physics, not once but time and again. Einstein discovered that light was not really a wave but a stream of discrete particles. He demonstrated that objects could not be accelerated indefinitely but would bump up against a fundamental limit of our universe: the absolute maximum speed of an object is the speed of light in a vacuum, 186,282 miles per second. He showed that time was relative, that clocks in different parts of the universe would tick at different rates. And it was Einstein and the mathematician Hermann Minkowski who, in defiance of everyone's senses, maintained that we live not in the three dimensions of Euclid but in four dimensions.

The possibility of four dimensions had been broached in the nineteenth century at the same time Euclid's parallel postulate was crumbling. Talk of parallel lines meeting had opened the floodgates to mind-blowing alternative geometries. If despite appearances to the contrary, the angles of triangles in the world don't really sum to 180 degrees, people wondered how they could trust their senses that they live in three dimensions and not, say, four. In 1884, Edwin Abbott Abbott, a brilliant theologian and Shakespearean scholar, anonymously published the second edition of *Flatland*, a literary *jeu d'esprit* that is at once a sly satire of the one-dimensional thinking of Victorians (their class distinctions and their view of women as inferior) and a glimpse of the fourth dimension by, paradoxically, imagining the life of beings trapped in two dimensions.

"I call our world Flatland," begins the narrator of Abbott's slim book, "not because we call it so, but to make its nature clearer to you, my happy readers, who are privileged to live in Space." Picture intelligent pancakelike aliens confined to a two-dimensional surface, confined not just physically but sensorily—these creatures have no faculties to sense anything "off" the surface. Imagine a sphere descending on Flatland and passing through it. What will the Flatlanders experience? They will not sense its approach or know its solidity. They'll experience first contact as a point, which will grow into an ever-expanding circle as the first half of the sphere cuts through Flatland and then contracts back to a point and vanishes as the trailing half of the sphere passes through. When a visitor from Spaceland (the three-dimensional world) tells the Flatlanders of evidence for a third dimension, they fly into a rage—as does the Spacelander himself when subsequently presented with evidence of a fourth dimension.

Erdős was once spotted at a party, hunched over on a couch, studying *Flatland*, the only work of fiction he may have read cover to cover as an adult. The idea of a fourth dimension, even after Einstein made a strong case for it, is not easy for our intuition to accept. In 1920, Sir Arthur Eddington, the brilliant physicist, confessed that

> however successful the theory of a four-dimensional world may be, it is difficult to ignore a voice inside us which whispers: "At the back of your mind, you know a fourth dimension is all nonsense." I fancy that voice must have had a busy time in the past history of physics. What nonsense to say that this solid table on which I am writing is a collection of electrons moving with pro-

digious speed in empty spaces, which relative to electronic dimensions are as wide as the spaces between the planets in the solar system! What nonsense to say that the thin air is trying to crush my body with a load of 14 lbs. to the square inch! What nonsense that the star cluster which I see through the telescope, obviously there *now*, is a glimpse into a past age 50,000 years ago! Let us not be beguiled by this voice. It is discredited. . . .

In his years at the Institute, from 1933 to his death in 1955, Einstein tried unsuccessfully to construct a unified theory of gravity and light that would treat both as different manifestations of a single phenomenon. He also worked on the foundations of quantum mechanics, the physics of the very small. Though he was the king of counterintuitiveness, he could never make his peace with certain paradoxical aspects of quantum mechanics, like the famous Heisenberg Uncertainty Principle, according to which the more accurately you know a subatomic particle's velocity, the less you know of its whereabouts. Carrying the principle to the extreme, if you know the velocity precisely, you cannot rule out the particle being anywhere in the universe. It was in this connection that Einstein invoked the deity, serving up his famous sound bite: "God does not play dice with the universe."

Einstein and Gödel were close friends. Like Erdős, Einstein tried to pull Gödel out of his crisis of confidence, and succeeded in turning him on to relativity theory for a sufficient time to produce one important paper. Mostly, though, Einstein tried to keep Gödel out of trouble. Gödel's paranoia led him to see contradictions not just in the foundations of mathematics but in other hallowed subjects as well. While reading the U.S. Constitution in preparation for his citizenship examination, Gödel became convinced that he had found an inconsistency that allowed for the possibility not of a president but of a dictator. Gödel was irate—he had come to America to avoid dictators like Mussolini and Hitler. During Gödel's oral examination for citizenship, Einstein had to restrain him (interrupting and cutting him off) from sharing with the examiner his appalling discovery.

Erdős spent time at the Institute with Gödel and Einstein, although he didn't collaborate with either. "They don't have Erdős number one," he said, smiling. "I was not close friends with Einstein, but I knew him quite well. At lunch once at his house, I explained to him the Prime Number Theorem. Of course, he understood it and said it was nice but he had little interest in the details." Einstein's physics, of course, was very mathematical, and he is responsible for the most famous equation of modern times, $E = mc^2$, which expresses the idea that energy and mass are equivalent (with $c$ being the speed of light). Most of Einstein and Erdős's conversations were about politics, which, to Einstein's dismay, occupied more of his time in the 1940s than physics did. "We have to divide up our time like that, between our politics and our equations," he told Ernst Straus, his mathematical assistant. "But to me our equations are much more important, for politics are only a matter of present concern. A mathematical equation stands forever."

"I think that the atomic bomb would have been done without Einstein," said Erdős. "He provided the basic understanding but it could have been done without the theory of relativity. I remember asking him in 1945, 'Did you think 40 years ago that your formula $E = mc^2$ would have any application in your lifetime?' And Einstein certainly said, 'No, I didn't. I expected that we would have an application eventually, but not so soon.'" The two men also discussed religion. "Einstein definitely did not believe in a personal God," said Erdős. "That I know because I asked him."

Erdős was much closer to Ernst Straus than he was to Einstein. Straus's widow, Louise, first met Paul Erdős at the Institute in 1944. "I remember that meeting well," she said, laughing, fifty-three years later. "Ernst left me alone with him while he went off to work with Einstein. Erdős, I quickly learned, had difficulty sitting still, particularly when he was doing mathematics. And as I looked after him, he was doing a lot of mathematics, which meant he was walking up and down the streets of Princeton, waving his hands, gesticulating wildly. Moving his hands apparently helped him visualize geometric subjects. I couldn't keep up with him, not with the mathematics nor with him physically. He walked quite fast. At some point I lost sight of him and couldn't find him. I was quite worried—I didn't know what my husband would think if I lost one of the greatest mathematical minds of the twentieth century— but I eventually found him. He was standing there, knocking his head against a building. He told me he had a headache, apparently from thinking so hard."

Louise and Ernst Straus were newlyweds, and they ended up in Princeton because Ernst had just started working for Einstein. The great physicist needed to verify things mathematically, and the Institute provided him with a stipend for an assistant. "After a lot of prodding by an acquaintance of Einstein, Ernst interviewed for the position," Louise Straus recalled. "The interview seemed to go well, but at one point Ernst felt compelled to say, 'I must tell you I don't know relativity theory.' 'That's all right,' Einstein replied, smiling, '*I* know relativity theory.' Ernst got the job. From 1944 to 1948, he worked for Einstein. We didn't have a telephone; Einstein would send over ideas by telegram—maybe the only time the details of relativity theory were telegraphed!" Mostly, though, Einstein was pursuing other subjects because he had dreamed up relativity forty years earlier. "Each morning my husband would stop by Einstein's house on Mercer Street," Louise Straus said, "and walk with him to his office at the Institute. Einstein had a large, beautiful office, with nice carpeting and a very elegant huge desk in a bay window. But he didn't like to work there. He preferred to work with Ernst in a tiny cubicle in the back. He said I could use the big office. So I'd sit at his big desk—I was finishing up my mathematics degree at Columbia at the time—but it didn't do me any good. Sometimes the head of the Institute would stop by to show distinguished guests Einstein at work, and there I'd be at his desk! Ernst and Einstein would work together until lunchtime. Then Einstein would go home. In the afternoon Ernst would finish the morning's calculations, or he'd do his own math, with other Institute guests like Paul Erdős.

"Erdős was always pacing, like he did the first time I met him. Years later, after we moved to California, we had an Irish setter that would follow him back and forth for hours whenever he paced, but Erdős was too engrossed to ever notice the dog. Once we were on a plane together, on a long trip, from Australia to Los Angeles. I can still picture him going up and down the aisle, waving his hands, discovering mathematical truths. People stared at him the whole flight. Then, when we got home, Ernst and I were tired and wanted to go to bed, but he was wide awake and wanted to continue doing mathematics all night.

"We lived in old army barracks at the Institute. Erdős was staying across the street in 'the bachelors' quarters,' with eight other people who shared a living room. Ernst once asked him who his roommates were. 'Trivial beings,' he replied. 'Trivial beings.' That meant they weren't mathematicians. When he got tired of the trivial beings, he'd just show up at our place, and we never knew how many days he was going to stay. I remember during the night hearing crashing sounds. The windows had no sash cords. If you opened the lock, they'd come crashing down. He was such an intelligent man but he could never figure out how to gently lower the windows. He was the real absent-minded professor. He couldn't figure out how to manage the shower. He could never shut the faucets off. Water ran out on the floor. The linoleum buckled, and the door wouldn't shut again. He'd go outside to the pay phone and drop coins in it all night, calling mathematicians around the world and asking nearby friends to come over to our place. 'I'm at the Straus house,' he'd tell them. He never asked us first if we wanted more guests. He'd just invite all the mathematicians over. But, I must say, my husband loved it. They knocked all sorts of ideas around. A lot of good mathematics came out of it. If Ernst were alive, he'd tell you what exciting times those were mathematically. It was a heady time at the Institute. Erdős was stirring up a lot of research, Einstein was always stimulating, and von Neumann was building the first computer.

"Erdős was staying with us in Princeton when my husband came down with diabetes in 1948. Paul wanted to take insulin, too. Of course, we wouldn't let him. He was furious. The scientific curiosity in him made him want to try it. Ernst and I were also learning to drive at the time. We wanted to drive cross country, so we bought an old car. Paul wanted to drive, too. We wouldn't let him. If he had driven, that would have been it. An accident would have ended it all. We had all sorts of arguments. He was very insistent.

"In the summer of 1948, just before we moved to California, we had a party. He always had trouble tying his shoes, and I remember him sticking his foot out at the party, asking people to tie his shoe. People joked that we were moving to California just to get away from him."

Ernst Straus was one of the few people who had the opportunity to observe firsthand the differences in style between the master physicist and the master mathematician. In a tribute to Erdős on his seventieth birthday, Straus said: "Einstein often told me that the reason he chose physics over mathematics was that mathematics is so full of beautiful and attractive questions that one might easily waste one's powers in pursuing them without finding the central questions. In physics he had the 'nose' for the central questions and he felt that it was the chief duty of the scientist to pursue those questions and not let himself be seduced by any problem—no matter how difficult or attractive it might be. Erdős has consistently and successfully violated every one of Einstein's prescriptions. He has succumbed to the seduction of every beautiful problem he has encountered—and a great number have succumbed to him. This just proves to me that in the search for truth there is room for Don Juans like Erdős and Sir Galahads like Einstein."

∎

On December 2, 1948, Erdős returned to Budapest, his first time back after a decade abroad. The trip was bittersweet: many of his friends and relatives were dead, but his mother and closest friend Paul Turán were alive and well. It was then that he met Turán's future wife Vera Sós and a five-year-old prodigy named Miklós Simonovits, both of whom would later rank among his closest collaborators. He had to cut short his visit home, though, when Stalin started sealing the borders and rounding up civilians for the notorious scripted trials.

In February 1949, Erdős slipped out of Hungary. For the next three years he went back and forth between England and the United States before landing a flexible deal at the University of Notre Dame in 1952. He had to teach only one class and was assigned an assistant who could take over on the spur of the moment should he have the urge to rush off and finish a proof with a collaborator. Although Erdős rejected organized religion, he didn't mind teaching at a Catholic school. "The only thing that bothered me," he joked, "was that there were too many 'plus signs.'" Notre Dame offered to make his appointment permanent on the same generous terms. "His friends urged him to accept," recalled Melvin Henriksen, who met Erdős on his fortieth birthday, when Erdős was telling anyone who would listen: "Death begins at forty."

"At age forty when he began moaning and groaning about the SF having one hand on his shoulder," recalled the widow of one of his colleagues, "I remonstrated him with, 'Paul, if you feel so badly at forty, how will you feel at fifty?' His immediate and sad response was 'Worse.'"

His friends pleaded with him. "We said, 'Paul, how much longer can you keep up a life of being a traveling mathematician?' Little did we suspect," Henriksen said, "that the answer was in excess of forty years." Erdős turned Notre Dame down, because he didn't want to be pinned down by the responsibilities of a permanent job.

In July, he wanted to call his mother in Hungary on her birthday. To his chagrin, his friends were afraid to let him use their phones to call a Communist country. America was in the grip of the Red Scare.

"Then my problems started with Sam and Joe," Erdős said. "I didn't want to return to Hungary because of Joe. In 1954, I was invited to an international mathematics conference in Amsterdam. Sam didn't want to give me a reentry permit. It was the McCarthy era. The immigration officials asked me all sorts of silly questions. 'Does your mother have great influence on the Hungarian government? Have you read Marx, Engels, or Stalin?' 'No,' I said. 'What do you think of Marx?' they pressed. 'I'm not competent to judge,' I said, 'but no doubt he was a great man.' The only intelligent question they asked was, 'Is it easy to leave Hungary?' They knew of course that the answer was no, but they wanted to see what I was going to say. 'No, it isn't easy,' I said. 'I'm not planning to visit Hungary now because I don't know whether they would let me back out. I'm planning to go only to England and Holland.' "

Although the immigration authorities didn't like Erdős's answer about Marx, it was his answer to another question that apparently troubled them the most. " 'Would you visit Hungary if you could be sure that you could leave again?' they asked. And I said, 'Of course, my mother is there and I have many friends there.' But during the McCarthy era you couldn't admit to wanting to visit a Communist country."

So they denied him a reentry visa, and he hired a lawyer to appeal the decision, only to be refused again. "No reason was ever given," recalled Henriksen, "but his lawyer was permitted to examine a portion of the Erdős file and found recorded the facts that he corresponded with a Chinese number theorist named Hua who had left his position at the University of Illinois to return to Red China in 1949 (a typical Erdős letter would have begun: Dear Hua, Let $p$ be an odd prime . . . ) and that he had blundered onto a radar installation in Long Island . . . while discussing mathematics with two other noncitizens." The authorities apparently feared that the letters to Hua, filled with impenetrable mathematical symbols, might be coded messages.

Erdős resigned his position at Notre Dame, forfeited his green card, and headed to Amsterdam. "To Erdős," said Henriksen, "being denied the right to travel was like being denied the right to breathe." He never wanted to be restricted. "I left without a reentry permit," Erdős said, "which I think was done in the best American tradition: that you don't let yourself be pushed around by the government. . . . For many years I couldn't return."

"I met Erdős in Europe after McCarthy kicked him out of America," recalled Anne Davenport, some forty years later. "In the taxi on the way home he wanted to find something. He had two suitcases with him. That was all he had. He opened both, and each was only a third full. They contained all of his worldly possessions, and none are worth mentioning. When we got home, he wanted me to call his mother in Budapest. I asked him what her number was. He said that she didn't have a phone. I asked him how he expected me to call. He said her neighbor had a phone. I asked him for the neighbor's number. 'I don't know,' he said. 'That's for you to find out.'"

Always an optimist, Erdős expected that Western European countries would be kinder than Sam in letting him travel unrestricted. But he met resistance there, too. Holland, to his disappointment, would give him a visa that was good for only a couple of months, and England wasn't any better; those countries too didn't like him fraternizing with Red Chinese. At last Israel bailed him out with a three-month appointment at Hebrew University in Jerusalem, and it was there that he had to pay out the first monetary reward, $20, for the solution to one of his challenges, a tantalizing problem in set theory. He became a resident of Israel, although he turned down an offer of Israeli citizenship and kept his Hungarian passport, declaring himself a citizen of the world.

In 1955, two years after Stalin's death, Erdős visited Hungary when well-connected friends, arguing that Erdős was a singular asset to the world culture of mathematics, convinced the government to grant him a special passport that said he was a Hungarian citizen but acknowledged his Israeli residency. (His mother may also have helped lobby the Hungarian government; although she was apolitical, the Communist regime had a long memory and regarded her warmly because she did not join the counterrevolutionary forces back in 1919 in the effort to topple Béla Kun.) The special passport—which was granted only to Erdős—allowed him to go in and out of Hungary whenever he pleased. In 1956, a pro-democracy revolution in Budapest was crushed by Soviet tanks. The new Soviet-installed dictator, János Kádár, who would rule Hungary for thirty-two years, reaffirmed Erdős's unusual passport.

Erdős had made his peace with Joe—and János, as he called Hungary after 1956. He could see his mother all he wanted and work again with his Hungarian colleagues. He and Turán would take breaks from their mathematical collaborations by rewriting classic Hungarian poems. "The central theme of their 'compositions,'" said János Pach, "was old age and senility, the two things that terrified them most." Erdős was fond of reciting one couplet in particular:

> One thought disturbs me, that I may decease
> In slowly progressing Alzheimer's disease.

"Erdős knew that he wasn't very good at remembering names," Graham said, "but he often said he'd know he really was in trouble when he forgot the name *Alzheimer's.*"

# 3

## EINSTEIN VS. DOSTOYEVSKY

My own greatest debt to Erdős arises from a conversation 30 years ago in the Hotel Parco del Principi in Rome. He came up and surprised me by saying, "Guy, veel you have a coffee?" I don't drink much coffee, but I was intrigued as to why the great man had singled me out. Coffees were a dollar each, about standard today, but then it seemed a small fortune. When we got our coffee, Paul said, "Guy, you are eenfeeneetely reech; lend me $100."

I was amazed, not so much at the request, but rather at my ability to satisfy it. Once again, Erdős knew me better than I know myself. Ever since then, I've realized that I'm infinitely rich: not just in the material sense that I have everything I need, but infinitely rich in spirit in having mathematics and having known Erdős.

—Richard Guy

In 1959, Erdős caught Sam in a good mood and was granted a special visa to attend a number theory conference in Boulder, Colorado. The visa required him to be accompanied by a fellow mathematician at all times—which was hardly a burden because that was the only company he ever kept—and to leave the country promptly after the conference.

"When I returned from the United States in the summer of 1959," recalled Erdős, "I was told that there is a little boy whose mother is a mathematician and who knows all there is to be known in high school. I was very interested and next day I had lunch with him. . . . While we had lunch and [Louis] Pósa was eating his soup, I told him the following problem. Prove that if you have $n + 1$ integers less than or equal to $2n$ there are always two of them which are relatively prime. I discovered this simple result some years ago but it took me about ten minutes until I found the very simple proof."

As an example, choose $n$ to be 5. Then the conjecture is that if you take any six integers from the set 1, 2, 3, 4, 5, 6, 7, 8, 9, and 10, you can't avoid choosing two that are relatively prime (meaning, remember, that they have no common divisor greater than 1). The conjecture would fail if you were allowed to choose just five of these integers: there are five even numbers in this set, namely, 2, 4, 6, 8, and 10, all of which obviously share the divisor 2.

Louis Pósa finished his soup and announced, "The two are neighbors." In other words, the two are consecutive. "If you have $n + 1$ integers less than or equal to $2n$," said Erdős, "two of them are consecutive and therefore they are relatively prime." Needless to say, Erdős was "very much impressed." After he told this story at a lecture, one wit commented, "On this occasion champagne would have been more appropriate than soup." Erdős thought that Pósa's soup proof demonstrated that the twelve-year-old was on the same level as the great Gauss, who at the age of ten quickly summed all the integers from 1 to 100.

Erdős took Pósa under his wing. They got together often, and Erdős's mother plied Pósa with cakes and sweet beverages. "When he was a little over thirteen," said Erdős, "I explained to him Ramsey's theorem," and posed a problem involving a graph with an infinite number of points and an infinite number of edges. Erdős challenged Pósa to find an infinite subset of this graph in which the points are all connected or all disconnected. This is just the infinite version of the party problem. Is there always either an infinite clique within the party all of whose members know each other or an infinite subgroup none of whom know each other? "It took about fifteen minutes until Pósa understood it," said Erdős, "and then he went home, thought about it all evening, and before going to sleep he had the proof.

"By the time Pósa was about fourteen you could talk to him as a grown-up mathematician. I called him on the phone and asked him about a problem. If the problem was about elementary mathematics, it was very likely that he had some relevant and intelligent comment. It is perhaps interesting to remark that he had some difficulty with calculus. . . . He never liked geometry. I tried to give him some math problems in elementary geometry but he never liked them. He always liked to do only what he was really interested in, but at that he was extremely good."

Pósa and Erdős wrote their first paper together when Pósa was fourteen. He did his most famous work, in graph theory, when he was fifteen. But by the age of twenty he had stopped proving and conjecturing, and was teaching grade school. "I often comment sadly," said Erdős, "that he is dead, but I very much hope that he will come back to life soon. I got first worried about him when he told me when he was sixteen that he rather would be Dostoyevsky than Einstein."

To the detriment of his mathematics, young Pósa was also distracted by girls. He asked Erdős why there were so few "girl mathematicians," and Erdős told him: "Suppose the slave children would be brought up with the idea that if they are very clever, the bosses will not like them. Would there be then many boys who do mathematics?" Pósa agreed that there wouldn't be very many.

If Erdős couldn't help him with girls, he could turn him on to other pleasures. "In Hungary many mathematicians drink strong coffee," said Erdős. "At the mathematical institute they make particularly good coffee. When Pósa was not quite fourteen, I offered him a little coffee, which he drank with an infinite amount of sugar. My mother was very angry that I gave the little boy strong coffee. I answered that Pósa could have said: 'Madam, I do a mathematician's work and drink a mathematician's drink.' I saw a movie many years ago where a lady sees a boy of sixteen drink whiskey with an older man and is shocked. The boy says: 'Madam, I do a man's work and drink a man's drink.' "

Pósa was one of the first students to attend a special high-school program for children gifted in mathematics. He liked the program so much that he stayed there even though his test scores qualified him to leave two years early to attend college. He told Erdős that "there are boys in my class who are better in elementary mathematics than I."

One of these boys was László Lovász, who would become famous for his work in combinatorics and would write seven papers with Erdős. Lovász only started doing serious mathematics late in life, said Erdős, "at the ripe old age of nearly seventeen. When Lovász was still an epsilon, in the first year of high school, he and . . . a fellow mathematician courted the same boss-child, also a mathematician and not a bad one as bosses go. The two slave children asked her to choose. She chose Lovász," and they got married. But the love story could be improved upon, Erdős noted, by having the boss-child answer: "I will choose the one who proves the Riemann Hypothesis."

Erdős made it his mission to seek out child prodigies all over the world. József Pelikán, who was fifteen when they met, said Erdős nurtured young talent by "pouring problems on us as if we were professional mathematicians." The attention paid off. Though a few of his young charges "died" prematurely like Pósa, many others grew up to be among the leading mathematicians of our time.

Erdős loved all children, not just the mathematically precocious, and doted on the epsilons of his collaborators. "People are always taking pictures of me holding babies," he said. In one photo, "the baby looked so content that somebody said, 'Uncle Paul is nursing.'" The younger the child was, the deeper his connection.

Aleksandar Ivić's two-year-old daughter, Natalija, once accompanied the two men to a Belgrade park, where they sat and tried to fathom the distribution of prime numbers. At some point Ivić remembered a quick chore he had to do and, at Erdős's request, with some trepidation left his daughter with him. When Ivić returned, the park was empty. "No Erdős, no Natalija," he recalled. "I felt sick in the stomach and looked around once, twice, three times.... Panic was starting to take over, and dark thoughts of death, kidnapping, traffic accidents came to my head." As he was about to call the police, he spotted them across the street. "They were walking slowly, Natalija holding Erdős's hand, and they seemed to be talking (in what language I never knew!) and smiling at each other." In Natalija's other hand was candy. The two of them looked perfectly content. "I ran toward them and hugged them both. 'You shouldn't have worried,' Erdős said with a sly smile, 'I have my ways with children.'"

His memory for children rivaled his memory for mathematics. "When he asked a father of four, 'How are the epsilons?', it did not mean that he did not remember the names of the children. Just on the contrary—he perfectly well knew the names, ages, past illnesses," and other significant events of those children and "a few thousand others," said Pelikán, now a graph theorist at Eötvös Loránd University in Budapest.

Erdős also forged a special bond with anyone he perceived as vulnerable. In 1945, when Michael Golomb was in Philadelphia at the Franklin Institute doing war service for the government, he got a call from Erdős, who was passing through town and wanted to get together. Golomb explained that he was going to a party that evening at the home of a fellow mathematician who was eager to meet Erdős and would be happy if he showed up. Erdős did come to the party, but instead of conversing with the mathematicians he promptly disappeared. "We did not see him for the rest of the evening," Golomb recalled. "Only when everyone was ready to leave did we learn that Erdős had found out that our host had a blind father, who could not join the party, but sat up in a room on the upper floor. Erdős preferred spending the time with the lonely blind man rather than with the people in the party, who were eager to meet him."

Peter Winkler saw the same side of Erdős when he was teaching at Emory. "We had a very bright student who had cerebral palsy and was in a wheelchair," said Winkler. "When Paul saw him, he immediately came over and asked him what was his disease and what was his prognosis. He found out more about this student in ten minutes than we had learned in the entire time he was a graduate student in the department. And then he got into what the student was doing—he was working on his Ph.D.—and made some suggestions. It was wonderful. This was the kind of thing he did all the time."

■

In the early 1960s, Erdős repeatedly petitioned the U.S. government to allow him reentry, lining up letters of support from university presidents, other bigwig academics, even U.S. senators. "His requests were rejected again and again," said Michael Golomb. "I received a letter from him (in 1961 or 1962) saying that he had at last obtained the promise of an American consul that he would receive the visa shortly, but a few weeks later, he had to cancel his visit; the promise was disavowed. In his typical style he wrote that the foreign policy of the State Department was adamant on two points: nonadmission of Red China to the United Nations and of Paul Erdős to the United States."

During the summer of 1963, Ernst Straus and John Selfridge, the American number theorist who had met Erdős ten years earlier, collected the signatures of a couple hundred mathematicians asking the State Department to readmit him. "He was able to return in November," said Selfridge, "just before the Kennedy assassination." "Sam finally admitted me," said Erdős, who was fifty at the time, "because he thinks I'm too old and decrepit now to overthrow him." (But his problems with governments were not over. He returned to Budapest at least once a year until 1973, when Kádár, under pressure from Moscow, banned Israeli mathematicians from a conference on set theory held in Keszthely, Hungary, in honor of Erdős's sixtieth birthday. Erdős was incensed and talked about boycotting his own birthday party. He ended up going, but then in protest refused to return until 1976, breaking his boycott only when Vera Sós summoned him to the bedside of Paul Turán, his childhood friend, who was dying of cancer.)

In 1964, his mother, at the age of eighty-four, started traveling with him. For the next seven years she accompanied him everywhere except to India, which she avoided because of her fear of disease. His mother disliked traveling—she barely knew a word of English, and Erdős traveled regularly to English-speaking countries—but she wanted to be with him. Wherever he did mathematics, she sat quietly, basking in his genius. They ate every meal together, and at night he held her hand until she fell asleep. "She saw in Paul the world," Cousin Fredro recalled. "He was her God, her everything. They stayed with me in 1968 or 1969. When they were together, I was nobody. It was like I didn't exist. That hurt me a lot, because I was very close to her. She was my aunt, and when I got out of Auschwitz, I went first to her home. She fed me and bathed me and clothed me and made me a human being again."

Erdős's mother worried constantly about his health and also fretted over his physical safety. "He was always active," said Vázsonyi. "He was always climbing up on walls and walking along them like a balance beam. And if he saw a hill, he'd have to run up it to see what was on the other side. Whenever he did this, you could see his mother's heart palpitating. She was afraid he'd disappear over the top!" She never gave up this worry. In the late 1960s, Erdős and his mother stayed with Vázsonyi and his wife in Manhattan Beach, California. "Our house was a couple of hundred yards from the water. There was an esplanade where you could walk. Erdős said he wanted to take a walk and his mother was totally opposed to this. She said that she didn't want him to go because she was afraid of what would happen if the water came up. This was totally absurd— the esplanade was twenty feet higher than the water."

Erdős defied her and went for the walk. "We were waiting for him to come back," recalled Vázsonyi, "and the whole time she was worried sick about the waves." She should have been worried about his sense of direction instead, because "he couldn't have been gone more than ten minutes when the phone rang. It was a lady who said she had a gentleman show up on her doorstep who said he was visiting us but was lost and didn't know how to find his way back. I said, 'Tell him to stand on the esplanade and look north and he will see me waving.' So I walked out and sure enough there's Erdős four or five blocks away. How he could have gotten lost on the esplanade, which ran straight with no forks, I'll never know."

In his adolescent and college years, Erdős's mother also protected him from other women. Vázsonyi remembers a time in the early 1930s in Budapest when he was chatting with his girlfriend and Erdős in a courtyard below Erdős's apartment. "She and Erdős were having a playful conversation, when suddenly I heard his mother shouting in an alarmed voice from a few stories above, 'Who is that woman?' She was very relieved to find out that she was my girlfriend."

Erdős never did have a girlfriend—or, for that matter, a boyfriend. "When he was in his seventies, he told me that he never had sex," said Vázsonyi. "He said he had problems on this score. I remember how he put it: 'The privilege of pleasure in dealing with women has not been given to me.'" Erdős explained to some of his friends that he had a physical abnormality that stood in the way of sexual pleasure. "He told me that when blood started flowing into his penis, it caused him great pain," said John Selfridge. "I don't think that he went to a sex doctor to get it fixed up. I think he went to some doctor years ago who explained his condition to him but didn't tell him that it was correctable. Obviously there are a lot of things that doctors can do nowadays that maybe they couldn't do twenty or thirty or forty years ago. But it wasn't really an issue for him. Mathematics was his first love. He never came on to women—and he never wanted to."

On some occasions when Erdős talked about his sexuality, he played down his physical problem. "It's a very complicated situation," Erdős told a journalist when he was seventy. "Basically I have a psychological abnormality. I cannot stand sexual pleasure. It's peculiar. You know, I have a basic character that I always wanted to be different from other people. It's very, very much ingrained. From a very early age I automatically resisted pressure to be like others."

He abhorred discussions of sex as much as he disliked the act itself. "In the 1940s," said Vázsonyi, "Gerhard Hochschild and I spent a lot of time chasing women and even more time talking about chasing women. We discovered that Erdős couldn't really stand that kind of talk. So we went out of our way to talk about women a lot. That really annoyed Erdős. 'Don't be trivial,' he'd say."

Other friends tormented him with pictures of naked women, which he hated to look at. "Once when Erdős wanted to play bridge," recalled Selfridge, "I took out a special deck that looked ordinary from the back but had seminude girls on the front. András Hajnal said Paul wouldn't play with the deck. I told Paul it was all we had and that he'd have to avoid looking at anything on the cards except the spots because I knew he was not going to like the rest. 'Vot?' he said, examining the cards. 'This is terrible!' But after the third or fourth hand, he proudly declared, 'You know, it is definitely possible to avoid looking at the cards except for the pips.'"

In the late 1940s, during the Chinese civil war, Erdős

took part in a food drive for the Communist Chinese. "I remember walking into a big room in Los Angeles, at UCLA, I think," said Vázsonyi, "and there was Erdős and all these people making packages of food. Some mischief-makers who knew of his disgust at naked women offered to make a $100 donation if he'd go with them to a bur-lesque show." To their astonishment, he immediately took them up on the offer. Afterwards, when they forked over the $100, he revealed the secret of his victory: "See! I tricked you, you trivial beings! I took off my glasses and did not see a thing!"

Erdős's eccentricities did not keep women from wanting to get naked with him, but they had to settle for a Platonic relationship. The most persistent of these women was a fellow mathematician, Josephine Bruening, and she was only briefly in the picture. "I call her 'the other woman,'" says Vázsonyi. "His first love, of course, was his mother." In 1962, when Vázsonyi was living in California, he got a call from Erdős after not hearing from him for some time. "It wasn't unusual to get a call after months of not hearing from him," said Vázsonyi, "although often he wouldn't even call first and just showed up unannounced." This had happened previously, "early one Sunday morning . . . when I heard a terrible racket at our door downstairs. Damn it, I thought, the newspaper boy is giving me trouble again with the Sunday paper. So I lean out the window to give him hell, and behold, it is Erdős banging on the door. 'Why didn't you call me on the phone?' I asked. 'Why would I do that?' he said." But this time he did call ahead. "He told me he was at UCLA and asked me to come see him. I went and we started to talk. He said, 'Vázsonyi, you don't have to drive me around anymore.' And he kept looking over his shoulder in an exaggerated sort of way. And I didn't understand the gesture. So I said, 'Erdős, what does this mean?' and I imitated his looking over his shoulder. And he said, 'It means that *she* drives me around.' So I looked behind him and there was this woman sitting there. She was Jo Bruening. That's the way he introduced me.

"That Sunday they showed up at our house. I no longer had to drive him around, which was a great convenience. The two of them were always together." Like Erdős, she could be dogmatic in her politics. "One day we were driving to Laguna Beach and stopped on the way to see the mission," said Vázsonyi. "Jo absolutely refused to go into the mission because she hated Catholics and there was an entrance fee. She said she's not going to support the Catholics. But Erdős had no hang-ups about this, so we went inside and he fed the pigeons with my daughter. Then we went on to Laguna and that's when the shit really hit the fan because they had only one room for Erdős. They didn't have two rooms for Jo and him. The manager suggested that they sleep in the same room. And Erdős got very disturbed, and said, 'That's impossible.' "

Another time a couple arranged for Erdős and Bruening to go camping with them in the High Sierras. When Erdős found out that he and Bruening would have to share a tent, he went ballistic. "I can't do that," he ranted, "because she has a cold!" When his friends observed that she seemed quite well, he blurted out, "I can't do that because she is a woman!" After months of following him around, Bruening told Vázsonyi's wife, Laura, that she was tired of being Erdős's chauffeur and was going to dump him. Soon she disappeared, and Erdős never mentioned her again.

But Erdős had no qualms about sharing a room with his mother. "I remember once in the 1960s," said Vázsonyi, "getting them a nice two-room suite at a hotel in Westwood. It was very clean and pretty, but his mother was dissatisfied. 'It's dusty,' she said, even though it wasn't. We couldn't figure out what was really bothering her. Erdős called the front desk and asked for a cot to be placed in the bedroom. As soon as they brought the cot up, her objections vanished. He obviously knew her protests about lack of cleanliness were a sham. She simply didn't want him to sleep in the other room. We had no idea."

In 1971, Erdős's mother died of a bleeding ulcer in Calgary, Canada, where Erdős was giving a lecture. Apparently, she had been misdiagnosed, and otherwise her life might have been saved. Soon afterward Erdős started taking a lot of pills, first antidepressants and then amphetamines. As one of Hungary's leading scientists, he had no trouble getting sympathetic Hungarian doctors to prescribe drugs. "I was very depressed," Erdős said, "and Paul Turán, an old friend, reminded me, 'A strong fortress is our mathematics.'" Erdős took the advice to heart and started putting in nineteen-hour days, churning out papers that would change the course of mathematical history. Still, math proved more of a sieve than a fortress. Never again could he bring himself to sleep in the apartment that he and his mother shared in Budapest; he used it only to house visitors and moved into a guest suite at the Hungarian Academy of Sciences.

For the rest of his life he'd bring his mother up at odd moments. "I was walking across a courtyard to breakfast at a conference," recalled Herb Wilf, a combinatorialist at the University of Pennsylvania, "and Erdős, who had just had breakfast, was walking in the opposite direction. When our paths crossed, I offered my customary greeting, 'Good morning, Paul. How are you today?' He stopped dead in his tracks. Out of respect and deference, I stopped too. We just stood there silently. He was taking my question very seriously, giving it the same consideration he would if I had asked him about the asymptotics of partition theory.

His whole life was spent thinking hard about serious mathematical questions, and he treated this one no differently. Finally, after much reflection, he said: 'Herbert, today I am very sad.' And I said, 'I am sorry to hear that. Why are you sad, Paul?' He said, 'I am sad because I miss my mother. She is dead, you know.' I said, 'I know that, Paul. I know her death was very sad for you and for many of us, too. But wasn't that about five years ago?' He said, 'Yes, it was. But I miss her very much.' We stood there silently for a few awkward moments and then went our separate ways."

π

————

# DR. WORST CASE

Dear Ron,

When Paul Erdős's mother died, someone told me that she was the last of his close living relatives. In the end I concluded that you are the closest to an intimate family that he leaves.

Did you ever hear the anecdote about the two philosophers discussing the possibility of intelligent life in other parts of the universe? One brought up the point that if there were intelligent life on other planets, they would as likely as not be more advanced intellectually than humans, and we could expect them to have visited Earth. But, he said, "Where is the evidence of their presence?" The other leaned over and whispered in his ear, "Shh! Here we call ourselves Hungarians." It is the rare person like Erdős who gives a point to such a story.

So I mourn the passing of this phenomenon, this self-sacrificing savant, this superannuated child-genius. And to you, who have invested more than any other person in smoothing his way in this rough society, I express my sympathy and my thanks for all you have done.

With my sincerest good wishes,

Gordon Raisbeck

It was twilight in Watchung, New Jersey, May 12, 1997, and Ron Graham was on his backyard trampoline, unwinding from a long day of mathematics. "I tried to get Paul's mother up on the trampoline once," Graham said, "but she refused. Paul got up, though, and plopped right down." Graham was dangling in midair, a few feet above the trampoline, suspended by twenty Bungee cords attached to a padded metal ring that fit snugly around his waist. Ten of the Bungee cords jutted off to his right, joining up with a mountain climbing rope hooked to the top of a tree. The other ten cords, on the left, joined up with a rope attached to the top of his house. "My goal is a double somersault with a double twist," said Graham. "I've never done it on a trampoline before. The Bungee cords, which would break any fall, may give me the courage. When you're older, you're not as strong. When you're not as strong, you're more likely to miss. When you miss, you are likely to get hurt. When you get hurt, it takes longer to heal."

Graham was bouncing up and down, the cords shooting him like a slingshot high into the air. But the setup doesn't look that safe. The base of the supporting tree is partially hollowed out, and the fasteners that connect the Bungee cords to the ring around his waist show signs of metal fatigue. Graham had to convince his handyman, a former high school math teacher, to jerry-rig the Bungee contraption. He obliged, but only after making Graham sign a waiver that exempted him from responsibility should the cords fail. This evening the cords hold, but they weren't put to the test by a double somersault with a double twist. (That would have to wait a few weeks until he was inspired by a team of Russian émigré circus performers, "who eke out a living trying to teach forward rolls to first graders." They got up on Graham's trampoline and did *triple* somersaults with double twists.) This evening he was content shooting up and down, telling the story of his life.

Graham was born on Halloween in 1935, in Taft, California, the son of a welder and a burner. "I remember my mother, the burner, explaining the distinction," said Graham. " 'Welders put things together, and burners take things apart.' " In search of opportunities to weld and burn, Graham's family moved every year or two, following jobs in oil fields and shipyards from California to Georgia to Florida. "We moved so many times I can't remember all the places we lived. Whole years of my childhood are blank. Back then," he said, softly, "I didn't always have the most fun."

It wasn't fun when his father skipped town when he was six, and Graham didn't see or hear from him again for another six years. Then one morning in Berkeley, California, when Graham was delivering newspapers, which he did twice a day to help his mother make ends meet, he looked up and saw a familiar man staring at him. "Excuse me," Graham said, "are you my father?" The man gave him a nifty pocketwatch and warned him not to tell his mother that he was in town. He didn't want her to come after him for money he owed her.

Navigating between his parents wasn't easy, and he didn't have many friends to turn to. "When you keep moving a lot," said Graham, "it's hard to make friends. I had only one date in high school. I drove her by a cemetery where her boyfriend had been buried three months before. She started crying. Needless to say, the date was not a great success."

Graham has always been a good athlete, although in grade school he was small for his age—"too small," he said, "to play popular sports like football. So I took up tumbling and juggling where being small was not a disadvantage."

(Today, owing to a late growth spurt in college, Graham is six feet two, gigantic for a gymnast.) As Graham criss-crossed the country, his gymnastics and juggling skills made him quick friends with fellow performers. "When you see someone juggling five balls, you know that he went through the same long learning curve that you did. That creates a common bond and instant camaraderie."

Compared to Erdős, Graham was a relative latecomer to mathematics, and even when he ultimately fell under its spell, the path he took to doing it professionally was quite circuitous. As a child, he had a good memory for the house numbers on his paper routes and had fun playing around with them in his head. A fifth-grade teacher, Miss Smith, showed Graham how to calculate square roots; he found that too easy, so he took it upon himself to extend the method to cube roots. At that age, however, he was more taken with stars and galaxies than with numbers; later he even considered becoming an astronomer, but abandoned the notion when he found out that astronomers spent more time staring at sheets of data than they did peering through telescopes.

His mathematical curiosity was piqued in seventh grade. "We were living then in a housing project in Richmond, California, which was a big shipbuilding community. Our place had chalkboard walls—you could put your hand through them. I was fortunate, though, to have a good math teacher, Richard Schwab, who was also my chess coach. I still have a photo of him," Graham said. He reached into a box of childhood memorabilia and pulled out a faded clipping from a 1947 newspaper. "This is Schwab," he said fondly, pointing to a dapper gentleman, deep in concentration, huddled over young Ronnie, who is playing black for Harry Ells Junior High School in the opening game of a match against the California School for

School portraits of Paul Erdős in Budapest— unusual because Erdős stayed home from school most years. His mother, Anna, feared that he might catch a fatal disease from other children.

A snapshot taken for Erdős's passport, which he would use in traveling to 25 countries.

C. 1921: Erdős, age 8, in one of his mother's favorite pictures.

LEFT: c. 1916: Erdős, age 3, with his mother at Lake Balaton in central Hungary.

BELOW: Decades later: Erdős and his mother relax in Mátraháza, at a guest house run by the Hungarian Academy of Sciences.

August 14, 1941: A bemused Erdős, age 28, and two fellow mathematicians, Arthur Stone and Shizuo Kakutani, after their interrogation by
the FBI. The three mathematicians were charged with trespassing and
suspicious loitering near a military radio transmitter on Long Island.
*Courtesy Daily News, L.P. Photo*

1955: Under Stalin and Lenin's portraits, Erdős talks with Hungarian elementary school children in Stalinváros ("Stalintown"), known today as Dunaujváros.

Erdős liked to lecture, or "preach," as he called it, and he was always at home in front of a blackboard.

Erdős, who put in 19-hour days proving and conjecturing, denied that he fell asleep during mathematics conferences. "I wasn't sleeping," he would say. "I was thinking."

RANDOM GRAPHS'89
Poznan, Poland
AUGUST 7-11, 1989

INSTITUTE OF MATHEMATICS
Adam Mickiewicz University

August 1989: Sleeping or thinking? Erdős is the only one with his head hung low in the formal group portrait at the Random Graph Conference in Poznań, Poland. (He is seated in the second row, second from the right. Graham is next to him, on the end. Joel Spencer is in front of Graham, and next to Spencer is Fan Chung.)

May 1995: At a reception at Emory University, new honorary-degree holders Erdős and Hank Aaron are joined by mathematician Ron Gould (standing) and an Emory official. On April 8, 1974, Aaron hit his 715th home run, topping Babe Ruth's 1935 record of 714. The numbers 714 and 715 have a special mathematical property—the sums of their respective prime factors are equal—a property Erdős succeeded in proving was true of infinitely many other pairs of numbers. *Courtesy Ron and Madelyn Gould.*

August 1989: At the Random Graph Conference Erdős supervises a "random" race. To begin the race, Erdős tosses a huge die. This determines the initial number of laps. But as the mathematicians approach the finish line, Erdős throws the die again, specifying an additional number of laps.

Erdős loved epsilons—his word for small children (in mathematics the Greek letter epsilon is used to represent small quantities). The photo of Erdős and the three epsilons was taken in the 1970s in California; the girl on the left, Bobbi, is Andrew Vázsonyi's daughter.

G. H. Hardy, the great
English number theorist,
whom Erdős met on his
second day outside
Hungary, abhorred
applied mathematics:
"No discovery of mine
has made, or is likely to
make, directly or indi-
rectly, for good or ill, the
least difference to the
amenity of the world."
*Courtesy The Master
and Fellows of Trinity
College, Cambridge*

1919 passport photo: Srinivasa
Ramanujan, the self-taught
Indian genius, was an inspira-
tion to Erdős, although the two
men never met.
"An equation for me," said
Ramanujan, "has no meaning
unless it expresses a thought of
God." *Courtesy The Master and
Fellows of Trinity College,
Cambridge*

Stanislaw Ulam (with wife Françoise), the Polish émigré mathematician, collaborated with Erdős over a 50-year period. "The first sign of senility," Ulam liked to say, "is that a man forgets his theorems, the second is that he forgets to zip up, the third sign is that he forgets to zip down." *Courtesy Los Alamos National Laboratory*

1993: George Szekeres and Esther Klein, both expatriot Hungarians, celebrated with Erdős and Chung at a conference in Keszthely, Hungary, in honor of Erdős's 80th birthday. In the early 1930s in Budapest, Szekeres, Klein, and Erdős had defied the Fascist government's prohibition on public meetings, by gathering daily in a park to talk politics and do mathematics.

1993: Vera Sós, one of Erdős's most prolific collaborators, also joined Erdős in Keszthely for his 80th birthday.

Sós waits for Erdős to have an insight into a problem. "Problems worthy of attack," Erdős believed, "prove their worth by fighting back."

Posing with two proto-mathematicians in Graham's kitchen: Erdős opened his brain to the famous as well as the unknown.

THE UNIVERSITY OF ADELAIDE

G.P.O., Box 498,
Adelaide,
South Australia 5001

Dear Fan + Ron (1985 IX 23)

I hope you had a pleasant time in Brazil. Selberg gave the Turán memorial talk in Budapest and left yesterday. I phoned Vera yesterday.

Please send 100 dollars in my name to the relief organisation of your choice for the earthquake in Mexico - the e.q. struck them with unearthly viciousness. Please send my mail between ℵ₁ and ℵ₂ to Univ of Western Ontario Math Dept London Ontario Canada c/o Prof D Borwein Math Dept

Kind regards

E. P.

---

August 9, 1985: A letter from Erdős noteworthy for the absence of math-ematical notation (unlike the five or so mathematical letters he wrote every day). Erdős, who gave away most of his money to charities, instructs Graham and Chung to contribute $100 to earthquake relief efforts.

1988: At a conference in Cambridge, England, Chung and Erdős taking questions after a lecture, and Graham and Erdős contemplating a proof.

Working with Chung and Graham on a nettlesome problem: Erdős more than anyone else was responsible for turning mathematics into a social activity.

Graham, past president of the International Jugglers Association, juggling 12 balls (though the world record is only 9), thanks to the digital retouching skills of his daughter, Ché. *Courtesy Ché Graham*

Graham relaxing on his Bungee trampoline after a long day at AT&T: "You can do mathematics anywhere. I once had a flash of insight into a stubborn problem in the middle of a back somersault with a triple twist."
*Courtesy Ché Graham*

Chung and Graham collaborating on their tandem bicycle near their home in New Jersey. "Many mathematicians," said Graham, "would hate to marry someone in the profession. They fear their relationship would be too competitive."
*Courtesy Ché Graham*

Graham, unlike Erdős, never forsook body for mind.
*Courtesy Ché Graham*

Mid 1980s: Erdős, oblivious to the camera, continues to be haunted by an intractable problem during a break at a combinatorics conference in the mountain resort of Hakone, Japan.

the Blind. ("And I thought, hey, my opponent's blind, it will be easy to beat him. Wrong!") Schwab did help Graham's chess—the position in the photo shows that Graham had mastered the first three moves in the slow-paced English opening—but did even more for his math.

"I knew algebra and trigonometry," said Graham, "and was confident I could solve any math problem thrown my way. But Schwab tried to stump me by posing this problem—and I remember it clearly now, fifty years later—about finding the size of a population of rats if you know that the rats die at a rate proportional to the size of the population. You'd like to know how large the population is at some specified time in the future. To solve the problem, Schwab said, you need calculus. You need to know about integration and differential equations. I had never heard of calculus, so I got hold of a calculus textbook and read it cover to cover. By the end of the semester, I could solve the rat problem. I was delighted with this world of mathematics. I was good at it, and mathematics, although not a complete refuge for me, was a world unto itself, a world that was clear, logical, and self-contained, a world that offered certainty." (*Certainty* is a word that mathematicians often use when they try to describe the appeal of what they do. "If anything is certain," Erdős once said, "it would be mathematics. It is absolutely certain.")

Graham enjoyed learning on his own. He ended up skipping a few grades and, at his mother's urging, won a Ford Foundation scholarship that took him at the age of fifteen, in 1951, to the University of Chicago. Chicago was a serious place; his classmates, all stars in high school, included Carl Sagan. "The university was in the midst of this educational experiment known as the Great Books program," recalled Graham. "You spent your undergraduate years reading Newton and Darwin in the original, reading

Plato, and reading Chaucer. I wasn't allowed to study math there because I had placed out of the subject on the entrance exam." He was enrolled in a club called Acrotheater, which combined dance, gymnastics, juggling, and other circus arts. "It was my favorite activity. We went around Chicago putting on elaborate acrobatic shows at local high schools."

After three years as an undergraduate, Graham's scholarship was running out. "It was not going to be renewed," said Graham. "I didn't have sterling grades. I cared more about juggling and the trampoline than I did about the Great Books. My father stepped in and offered to contribute to my tuition if I switched from 'the dangerous, leftist' University of Chicago to 'an all-American school' he knew like the University of California at Berkeley." Graham took his father up on the offer and enrolled in Berkeley. But again he had to put his mathematical ambitions on hold. He wasn't allowed to enter as a math major because he hadn't completed a formal course in calculus. So he majored in electrical engineering, while he caught up in math and moonlighted as an acrobat.

At Berkeley, Graham was fortunate to learn number theory from Derrick H. Lehmer, who along with his legendary father, Derrick N. Lehmer, had gone down in the mathematical record books for bagging large prime numbers. "During the second semester," Graham said, "he taught us about the proof of the Prime Number Theorem, which hadn't been out all that long. That was when I first learned about Erdős." After one year at Berkeley, however, Graham was worried about the draft. "I wasn't eligible for a student deferment because they wouldn't give one to someone who had been in college for four years with no degree. I didn't have anyone to turn to for advice. I saw this brochure for the air force, and decided to sign up. We weren't at war at the time, and I thought I'd have more control if I signed up instead of waiting to be drafted. I wanted to become an interpreter or a translator and study Russian or Chinese for a few years. It was said that if you scored first on the test—and I did score first—you could pick your assignment. You could pick anywhere in the world. But it turns out that is not the way it works. It turns out that you have to know the right person in the division. I didn't know anyone there so I was sent—this was in 1956—to Alaska, to Eielson Air Force Base about twenty-six miles south of Fairbanks, which was a better place to be sent, I suppose, than Thule, Greenland.

"I was assigned to be a communications specialist, really a telephone operator. I worked nights in the Service Club. Airmen would come in, usually after payday, have a few drinks, and call their families and girlfriends in the States. Alaska wasn't a state then, and sometimes it would take two hours to get an open line back to the mainland. A phone call cost seven dollars a minute. Nobody else wanted to work nights, but I liked it because the days were then my own." One of the things he did with his days was to attend the University of Alaska. But once again his plans for a formal mathematics education were thwarted; he had to settle for courses in physics because the university was not accredited in math.

"I was kind of lonely in Alaska, and I didn't like living in the barracks. One way you got to live off base was to get married. I knew this woman—Leah was her name—in Chicago. She was an acrobat. We got along. We were pretty close, although I didn't know her in the biblical sense. So I called her up one day and said, 'Hey, let's get married.' She said, 'You're kidding.' I said, 'No, why not?'

It turns out she was engaged. Also, I was naive about what you had to do to get married—the license, the ceremony, the relatives. But we did get married, in 1957. She came to Alaska, and we moved into the basement of a dance studio in Fairbanks only six miles from the university." By 1958, though, they had drifted apart. With a degree at last in physics from the University of Alaska, Graham persuaded the air force to let him serve out the last few months of his four-year tour of duty at a base in Sacramento. From there he returned to Berkeley in January 1959.

"I still was missing a lot of the math basics," said Graham, "but I almost had a thesis. It took me a while to catch on to the system." He signed up for S. S. Chern's class on differentiable manifolds, "and on the first day things were moving pretty fast. And the next day, people in the front rows were saying, 'Come on, let's get on to the good stuff!' Chern thought he was moving too slowly for the class, so he really started ripping. It turned out that graduate students typically audited a course once or twice before taking it for credit." Graham, who was seeing the course material for the first time, had to work extra hard. At Berkeley, Graham met Nancy Young, a fellow math student who would become his second wife. "As Nancy liked to say, we only took one mathematics course together and she got an A and I got a B." To earn money he formed a trampoline troupe, the Bouncing Baers, and a juggling comedy act, the Fumbling Franklins, which performed everywhere from two-bit store openings to circuses. ("We backed into the comedy. My partner wasn't a great juggler, so we had to add humor to our act.") In 1960, he tied for first place in the California collegiate trampoline championship. "I love the trampoline," said Graham. "It's a way of letting go. But there are limits to how far I'll let myself go. My 1960 co-champion leapt out of a plane with no parachute, falling freely and then grabbing a chute from someone who had jumped before him. California prosecuted him under its anti-suicide statutes."

∎

Graham's doctoral thesis was on unit fractions. These are fractions that are the reciprocals of positive integers, like 1/5, 1/8, and 1/127, fractions in which the numerator is 1. Such fractions were prized by the ancient Egyptians, who refused to deal with any fractions that weren't unit fractions (with the sole exception of 2/3, which had its own special hieroglyph). Most of what we know about Egyptian mathematics comes from papyrus rolls that have survived more than three and a half millennia. The most famous is the eighteen-foot-long, one-foot-wide Rhind or Ahmes papyrus from 1650 B.C. (Ahmes was the scribe who copied it from even older documents and "promised insights into all that exists, knowledge of all obscure secrets," and Henry Rhind was the Scottish antiquary who purchased it in the Nile resort of Luxor in 1858.) The papyrus starts with a table representing $2/n$ as a sum of distinct unit fractions for all odd values of $n$ from 5 to 101. For instance,

$$2/5 = 1/3 + 1/15$$
$$2/7 = 1/4 + 1/28$$
$$2/13 = 1/8 + 1/52 + 1/104$$
$$2/15 = 1/10 + 1/30$$

Why the Egyptians insisted on representing fractions this way is not at all clear. They may have found them simpler. Sometimes representing fractions this way makes it easier to tell when one is bigger than another. Is 55/84 larger than 7/11? The unit-fraction representation makes it clear that it is:

$$55/84 = 1/2 + 1/7 + 1/84$$
$$7/11 = 1/2 + 1/8 + 1/88$$

"I once asked André Weil, the legendary mathematician who is also a superb historian, why the Egyptians did this," said Graham. "And he said, 'They took a wrong turn.' I'm glad I didn't ask him this while I was in the middle of my thesis. That would have been discouraging!" The Greeks also took the wrong turn (and took another one with Roman numerals), and well into the seventeenth century unit fractions persisted throughout Europe as the preferred way of representing fractions. In his *Natural History*, Pliny the Elder estimated Europe's area to be "somewhat more than the third and eighth of the whole earth"—in other words, more than $1/3 + 1/8$, or $11/24$.

It is trivial to represent a fraction as a sum of unit fractions if you're allowed to repeat terms. The fraction 4/5, for instance, becomes $1/5 + 1/5 + 1/5 + 1/5$. But if we require all the denominators to be distinct, as the Egyptians did, is a unit representation always possible? It wasn't until 1202 that Leonardo Fibonacci, the greatest European mathematician of the Middle Ages, proved that the answer was yes.

It turns out in fact that any ordinary fraction can be expressed as a sum of unit fractions in *infinitely* many different ways. Using the identity $1/a = 1/(a + 1) + 1/a(a + 1)$, which Fibonacci knew, a sum of unit fractions can be continuously expanded. For example, $1/2 = 1/(2 + 1) + 1/2(2 + 1) = 1/3 + 1/6$. Applying the identity again to 1/3 expands the expansion: $1/2 = 1/4 + 1/12 + 1/6$.

And applying it still one more time to $1/4$ generates $1/2 = 1/5 + 1/20 + 1/12 + 1/6$. You can do this to your heart's content.

Fibonacci preferred to construct unit fractions by the so-called greedy procedure, in which the largest possible unit fraction is chosen for each term of the expansion. The "greedy" expansion of $3/7$, for instance, yields

$$3/7 = 1/3 + 1/11 + 1/231$$

(For those who must know the details, you get this expansion by taking the largest unit fraction less than $3/7$—namely, $1/3$, since $1/2$ is a shade bigger than $3/7$—and subtract the $1/3$ from $3/7$ to get $2/21$. Now the largest unit fraction less than the remainder of $2/21$ is $1/11$. Subtracting $1/11$ from $2/21$ leaves, *voilà*, $1/231$. Fibonacci showed that this greedy procedure always produces a sum of fractions that terminates.)

But no procedure, greedy or otherwise, is known for producing the best possible expansion, where "best" means that either the size of the largest denominator or the number of terms is minimized. In the expansion of $3/7$, the largest denominator can be held to 21:

$$3/7 = 1/6 + 1/7 + 1/14 + 1/21$$

Alternatively, the number of terms can be held to 3. For $3/7$, the greedy procedure happens to hold the number of terms to 3 (namely, $1/3 + 1/11 + 1/231$), but there is a "better" three-term expansion in which the largest denominator is not a whopping 231 but a much smaller 28:

$$3/7 = 1/4 + 1/7 + 1/28$$

For other fractions, the greedy procedure occasionally delivers neither the fewest number of terms nor the smallest denominator. Although the fraction 5/121 can be efficiently represented with just three terms

$$5/121 = 1/25 + 1/759 + 1/208,725$$

the greedy expansion yields the unwieldy

$$5/121 = 1/25 + 1/757 + 1/763,308 +$$
$$1/873,960,180,913 +$$
$$1/7,638,092,437,828,241,151,744$$

In 1955, Graham's instructor, Lehmer, who was worried about holding Graham's interest in class, challenged him with a problem that Herb Wilf posted in the *American Mathematical Monthly*: Prove that any ordinary fraction with an *odd* number for a denominator can be represented as the sum of unit fractions with *odd* numbers for denominators. For example, 2/7 can be represented as

$$2/7 = 1/5 + 1/13 + 1/115 + 1/10,465$$

Graham did the problem one better by thinking about which fractions could be represented as the sum of unit fractions with perfect squares for denominators. Thus, 1/3 can be expanded to

$$1/3 = 1/2^2 + 1/4^2 + 1/7^2 + 1/54^2 + 1/112^2 +$$
$$1/640^2 + 1/4,302^2 + 1/10,080^2 +$$
$$1/24,192^2 + 1/40,320^2 + 1/120,960^2$$

Graham proved that infinitely many fractions within a certain range could be represented by perfect squares.

"Lehmer sent my proof to an expert on unit fractions named Sherman Stein," said Graham. "Stein wrote him back that the proof was 'either the work of a very bright graduate student or a brilliant undergraduate.' That was very satisfying, but I was off to the air force."

When Graham got out of the air force in 1959, he wanted to make sure that his perfect squares conjecture really was original, so he sent it to the one man he had heard of who kept track of the mathematical literature in his head—Paul Erdős—who himself had written about unit fractions in 1932. "Erdős wrote back," said Graham, "saying that he hadn't heard of the result. I don't remember his response exactly but I can imagine that he pushed me further. That's what he always did. 'Have you thought not just about perfect squares but about higher powers?' he'd ask. 'Which fractions can be written as the reciprocals of cubes?'"

Erdős's 1932 paper is representative of how mathematical progress works in number theory, how specific results are made increasingly general. In 1915, a man named Taeisinger proved that if you add the reciprocals of the first $n$ numbers, you never get an integer. In other words, $1/1 + 1/2 + 1/3 + 1/4 + \ldots + 1/n$ never sums to an integer. In 1918, a man named Kurshchak dropped the requirement that the sequence start with 1. Add the reciprocals of any $n$ consecutive numbers and you won't get an integer. In 1932, Erdős discarded the requirement that the numbers be consecutive. They need only be uniformly spaced. Then their reciprocals still won't sum to an integer. For example, 1, 4, 7, 10, 13 have a common difference of 3. Erdős's proof rules out the possibility that $1/1 + 1/4 + 1/7 + 1/10 + 1/13$ sums to an integer and shows that it never will no matter how many more terms are added to the progression.

■

The name of Fibonacci, master of unit fractions, is eponymous in the mathematical literature not for the greedy procedure but for a silly problem he posed about rabbits: "A
certain man put a pair of rabbits in a place surrounded by a
wall. How many pairs of rabbits can be produced from that
pair in a year if it is supposed that every month each pair be-
gets a new pair which from the second month on becomes
productive?" At the start, there is one pair. By the end of the
first month, the first pair has begot another pair, so there are
two pairs total. By the end of the second month, the first pair
has produced another pair, for a total of three pairs. By the
end of the third month, the two pairs of the first month have
begot two more pairs, for a total of five pairs. The resulting
pairs form the sequence 1, 2, 3, 5, 8, 13, 21, 34, 55, 89, 144,
233, 377..., in which each number is the sum of the two
previous numbers. Fibonacci did nothing more than pose the
rabbit problem and give the answer of 377. He didn't study
the sequence of numbers that generated 377. That would
have to wait six centuries until François Édouard Anatole
Lucas, a French number theorist, investigated sequences that
started with any two positive integers and whose subsequent
numbers were the sum of the preceding two. The rabbit sequence was only the simplest such sequence, and the numbers in it came to be called *Fibonacci numbers*.

There are primes, like 2, 3, 5, and 13, among the Fibonacci numbers, but to this day no one knows whether
these primes are finite or infinite in number. It is easy to
choose other starting pairs for a Fibonacci-like sequence so
as to generate no prime numbers at all. Just choose as your
starting pair, say, 4 and 6. Now the Fibonacci-like sequence
4, 6, 10, 16, 26, 42, 68..., with all the terms divisible by

2, obviously doesn't generate any primes. Nor will any sequence that starts with two composite numbers that have a common divisor, say 10 and 15, which share the divisor 5; the sequence 10, 15, 25, 40, 65, 105 . . . can't include primes because the numbers are obviously all multiples of 5. Those are the trivial cases. Graham wondered whether he could find a prime-free Fibonacci-like sequence for the hard case of two starting numbers that are *relatively prime* (i.e., have no common factor larger than 1). He proved that such sequences exist, with the *smallest* known starting point being a pair of sixteen-and seventeen-digit numbers:

3,794,765,361,567,513 and 20,615,674,205,555,510

■

In 1962, after seven years in college, Graham finally got a degree in mathematics. "It never occurred to me that once I had my degree I would have to go to work. I had been in school all my life. I never thought about what I was going to do after school." And he still didn't have to give it much thought because he soon met David Slepian, a mathematician and recruiter from AT&T Bell Labs, who offered him a job in Murray Hill, New Jersey. It wasn't Graham's stint as a nighttime phone operator that made Slepian think Graham was right for AT&T; nor was it the reference Graham naively listed on his résumé—his gym coach. Rather, it was his impressive thesis on Egyptian fractions, which showed talent in the kind of mathematics (combinatorial number theory) on which the phone company depends. "My friends at Berkeley said I'd die mathematically if I took a corporate job instead of a university position," Graham said, "but I thought I'd try it for a few years, save a little money, and then figure out what I really wanted to do."

Thirty-six years later, Graham is still with AT&T in New Jersey. Nancy, too, has stayed in the AT&T fold, as a high-level systems programmer at Lucent Technologies, a spin-off of AT&T, but Nancy and he split up in 1978, after nineteen years of marriage. They have two grown kids, Marc and Ché, who have stayed clear of mathematics for careers respectively in software testing and digital imaging. They've dabbled in juggling, however, encouraged no doubt by the monetary awards their father offered. "I told them I'd give them $25 if they could juggle three balls for twenty-five throws, $100 for four balls, and $1,000 for five balls. 'What about six balls?' they asked. I said, 'No way. If you're crazy enough to master five balls, you'll probably learn to do six.' " They each won $100, as did the two kids of Fan Chung, whom Graham married in 1983.

Chung, like Graham, is one of the world's leading innovators in combinatorics. Graduating near the top of her class at Taiwan National University, she emigrated from Taiwan in 1970. "She entered the University of Pennsylvania as a graduate student," said Herb Wilf, who became her adviser. "I never paid any attention to the graduate students until they got past their qualifying exams. My policy then was to go after the best student and try to get him to go into combinatorics. The year she took the exam, 1971, she had the highest score by far—there was a huge gap between her and the next best student. So I immediately sought her out—I had never spoken to her before—and asked her if she knew anything about combinatorics. She said she knew a little from her days at Taiwan National University but not too much. I pulled out one of my magnetic subjects, Ramsey theory, that is guaranteed to get graduate students hooked on combinatorics because it is very pretty stuff. I gave her a book and told her to read the chapter on Ramsey theory. We set up an appointment in a week to talk about it. When she came to the appointment, I asked her how she liked the chapter. She smiled and said it was fine. Then she flipped the book open to a key theorem and said, gently, 'I think I can do a little better with the proof.' My eyes were bulging. I was very excited. I asked her to go to the blackboard and show me. What she wrote was incredible! In just one week, from a cold start, she had a major result in Ramsey theory. I told her she had just done two-thirds of a doctoral dissertation. 'Really?' she said softly. In fact, the result did become a major part of her dissertation."

"Many mathematicians," said Graham, "would hate to marry someone in the profession. They fear their relationship would be too competitive. In our case, not only are we both mathematicians, we both do work in the same areas. So we can understand and appreciate what the other is working on, and we can work on things together—and sometimes make good progress." Sometimes is an understatement; the two have published more than fifty important papers together and co-authored a book on Erdős-inspired results in graph theory.

■

When Chung and I first met, she was the head of mathematical research at Bellcore, a spin-off of Bell Labs that supported the regional phone companies after the breakup of AT&T. One of my friends from college, a mathematician-juggler like Graham, happened to work for Chung. I hadn't seen him in awhile, so I asked her how he was doing. "Stuart is great." She beamed. "Today he saved the phone companies millions of dollars without even knowing it!" Without even knowing it? How could that be? Harvard B.A., Stanford M.S., Columbia Ph.D., Stuart was not a dumb guy.

Discoveries in pure mathematics, which can seem unconnected to the real world, often turn out to have very practical applications, far from the minds of those who made them. That was the case with Stuart's work. Historically, many of the brightest minds in mathematics have prided themselves on doing math that has no applications. Math for math's sake was the rallying cry. They feared real-world relevance might distract from the pristine order and beauty that mathematics laid bare. When Euclid was investigating prime numbers, he was proud that they contributed nothing practical to Greek life. G. H. Hardy, too, reveled in his uselessness. "I have never done anything 'useful,'" he once said, not as an apology, but in defiance. "No discovery of mine has made, or is likely to make, directly or indirectly, for good or ill, the least difference to the amenity of the world." Hardy was a committed pacifist, who proudly claimed that his area of expertise, number theory, would never be used by the military. But he was wrong. In the past two decades, prime numbers, so thoroughly useless for 2,300 years, have found a home in the Pentagon as the basis of the military's most secure codes.

"This is the remarkable paradox of mathematics," observed commentator John Tierney. "No matter how determinedly its practitioners ignore the world, they consistently produce the best tools for understanding it. The Greeks decide to study, for no good reason, a curve called an ellipse, and 2,000 years later astronomers discover that it describes the way the planets move around the sun. Again, for no good reason, in 1854 a German mathematician,

Bernhard Riemann, wonders what would happen if he dis-
cards one of the hallowed postulates of Euclid's plane geometry. He builds a seemingly ridiculous assumption that
it's not possible to draw two lines parallel to each other.
His non-Euclidean geometry replaces Euclid's plane with a
bizarre abstraction called curved space, and then, 60 years
later, Einstein announces that this is the shape of the universe."

Unlike Hardy, Erdős was more accepting of the applications of mathematics, even though such applications did
not drive his own work. And he was realistic that in mathematics, as in the other sciences, "everything that can be
used for good things can also be used for bad things. After
all, the same differential equations which govern the spread
of poison gases also govern how pollutants spread. So one
can spread poison gases deliberately, but also one can prevent spreading of pollution."

But when the interests of Erdős's colleagues drifted
away from pure mathematics, he made no secret of his
disapproval. "When I wasn't sure whether to stay a
mathematician or go to the Technical University and become an engineer," Vázsonyi recalled, "Erdős warned me:
'I'll hide, and when you enter the Technical University, I
will shoot you.' That settled the matter." When probability theorist Mark Kac had a paper published in the *Journal of Applied Physics* based on his work during the war
at MIT's Radiation Laboratory, Erdős sent him a one-
sentence postcard: "I am praying for your soul." Erdős
was "reminding me," Kac said, "that I might be straying
from the path of true virtue, which, as a matter of fact,
I was."

Today, the distinction between pure and applied mathematics is more muddled than ever. The invention of the
computer has fueled the development of many fields of

mathematics, not because it is easier to do math on a computer (although sometimes it is) but because the inner workings of the computer are essentially mathematical. And physics, which has always borrowed the language of mathematics to describe the laws of nature, is now returning the favor by coming up with bizarre new physical objects, like the "superstrings" of modern cosmology, that demand entirely new mathematics.

In corporate research labs, like Bellcore and AT&T, the answers to very real-world questions may require the development of new mathematical ideas. Graham likes to cite an example from the days when AT&T was setting up private phone networks for corporate customers that had multiple locations. AT&T's tariffs were heavily regulated, and the government decided that the charge for a private network should be based not on how calls were actually routed through AT&T's wires but on the (theoretical) minimum length of lines needed to connect the different locations. One customer—Delta Air Lines, Graham thinks it was—had three main locations that were spaced equally, say 1,000 miles, apart. In other words, the three locations formed the vertices of an equilateral triangle. AT&T billed Delta for 2,000 miles of connecting wire because the shortest way to join them seemed obvious:

But Delta challenged the tariff, pointing out that if they opened a fourth office smack in the middle of the triangle formed by the other three, the total length of the connecting lines would be reduced by about 13.4 percent,

to 1,720 miles. Delta had rediscovered a mathematical observation first made in 1640 and pursued in the nineteenth
century by Swiss mathematician Jacob Steiner:

   The AT&T management went into a frenzy. What if
Delta claimed a fifth office? Could the connecting lines be
shortened even further? What if all its private-line customers started adding ghost offices to reduce the length of the
connecting lines? How much money would AT&T be out?
Graham was put on the case. In 1968, two of his Bell Labs
colleagues conjectured but could not prove that no matter
the size of the network, the addition of points would never
save more than 13.4 percent. Graham thought about the
problem on and off for years, and in the best tradition of
Erdős, offered $500 for a proof. He didn't have to pay up
until 1990, when Frank Hwang, one of his colleagues at
Bell Labs, and Ding Zhu Du, a postdoctoral student at
Princeton, delivered the proof.

■

On the wall of Graham's office at AT&T is a photo his
daughter Ché took of him juggling twelve balls. The world
record is only nine balls. "It was set by a man," said Graham, "who gave up juggling and went into the lawn-care
business." Graham can juggle seven balls erratically and
six consistently, but Ché digitally retouched the photo so
that her father has a dozen balls in the air. The retouching
is convincing: The balls cast little shadows on a brick wall

behind Graham, and the shadows are blurriest where the balls are moving the fastest, at the bottom of their arc, just before he's about to catch them. "I see juggling as a metaphor for many things in my life," said Graham. "Whenever you master a new trick, you can always add another ball. Whenever you've proved a new theorem in math, there's always another conjecture waiting to be proved. Erdős would never let you rest on the laurels of a proof but would point you to the next conjecture."

As cosmic justice has it, the chronically overscheduled Graham made his mathematical mark at AT&T on problems involving scheduling and efficiency. In the mid-1960s, Bell Labs was helping the government evaluate the Nike missile defense system. "The idea—and I think I just got the sanitized version—is that you have to defend against a lot of missiles coming in at you from all over the place. You have to know where they're coming from, what kind of missiles they are, who owns them—and you have to process this information fairly quickly because you might not get a second chance. You have a bunch of machines, a bunch of computer processors that are doing the analysis. The machines are pretty stupid and operate by simple rules: once a machine starts a task, it must finish it; if something more urgent comes along, you can't get a machine to drop a task in the middle. And if a machine is idle and *could* be doing something, it must be doing something. You'd think that if you added more machines, you'd definitely be able to speed up the analysis so that you'd finish sooner, but that's not the case. There are situations where if you add machines, it actually takes longer and you don't finish the job as soon. That's not a happy situation: *Kaboom!* The Defense Department wanted some guarantee that no matter what, the machines wouldn't take longer than whatever." It was Graham's job to determine just how bad it could be. He proved among other things that if you added machines, it can never get worse than twice as long.

Graham did a similar worst-case analysis for NASA. In the Apollo Moon missions, said Graham, "they had these three machines called astronauts." They had to perform dozens of tasks, from eating, sleeping, and exercising to conducting as many scientific experiments as time allowed. The tasks were so numerous that the optimal schedule was far from obvious. NASA set a schedule but worried about how much it might be improved. The space agency officials slept better at night once Graham was able to prove that the schedule it had established could not be improved upon by more than a few percentage points.

The Defense Department and NASA wanted a surefire method for finding the optimal schedules. In mathematics, such a surefire method is called an *algorithm*, a corruption of the last name of the ninth-century Persian number cruncher, Abu Ja'far Mohammed ibn Musa al-Khowarizmi, whose semantic legacy also includes the word *algebra*. An algorithm is simply a step-by-step procedure (like the greedy procedure for creating unit fractions) in which every step is explicitly stated so that a problem can be blindly solved by man or machine. What makes the Nike and NASA scheduling problems so challenging is that there is no known algorithm—other than the unenlightened, brute-force approach of enumerating and checking every possibility—for identifying the best schedule. And if the number of tasks that needs to be scheduled is sufficiently large, brute-force trial and error won't cut it. Even Deep Blue, the wily IBM computer that defeated Gary Kasparov by examining 200 million chess positions per second, would be stymied by the myriad possibilities.

The most famous computational problem that hasn't yielded to a clever algorithm is the traveling salesman problem: Given a network of cities and roads, find the shortest route that takes a salesman to every city exactly once. For a small set of cities and roads, it may be easy to find the solution, because there aren't that many roads to inspect. Even for a large network of cities and roads, you might get lucky and stumble on the optimal itinerary. But the only known method that always guarantees the optimal solution amounts to an inefficient, exhaustive search of all routes.

In 1971, Stephen Cook at the University of Toronto and Leonid Levin in the Soviet Union independently demonstrated that the traveling salesman problem, NASA's scheduling dilemma, and hundreds of other seemingly intractable computational problems are all equivalent mathematically in the sense that if an efficient algorithm exists for solving any one of them, all of them can be similarly solved. But most mathematicians believe—although they haven't been able to prove—that these problems will never yield to an efficient algorithm. The search for a proof of these problems' inherent intractability is one of the holy grails of the branch of mathematics known as complexity theory.

With no efficient algorithms in sight, complexity theorists have developed methods for finding less than optimal solutions, but solutions nonetheless. Thus they can send Willy Loman on his way, even if in theory he could take a shorter route. Graham's employer, while not in the business of dispatching salesmen to sell telephones door to door, has many computational challenges that are equivalent to the traveling salesman problem. There's the problem of efficiently routing a phone call through a network, and the problem of economically hooking up a host of elements on a microchip. Ronald Graham, in his early work at Bell Labs, pioneered techniques for determining how far off a given solution is from the theoretically optimal one; so though he can't tell you the best answer, he can tell you by how much your answer falls short of the optimal. In his personal life Graham is unflaggingly optimistic, but he has secured his place in mathematical history as the father of "worst-case analysis."

The worst case for some problems is not always that bad. Consider the problem of dividing a group of weights into two piles that are as close as possible in total weight. Like the traveling salesman problem, this simply worded problem is immune to fast algorithms, although it turns out to yield to a strategy that finds a good but not the best answer. Consider the simple case of five weights, three of which weigh 2 pounds each and two of which weigh 3 pounds each. In this simple case it's easy to find the exact answer by inspecting all the possibilities; the solution is to place the three 2-pound weights in one pile and the two 3-pound weights in the other, so that the piles are of equal weight:

3, 3                    2, 2, 2

But increase the number of weights dramatically, and trial and error is out of the question. In those cases, Graham suggests applying the following algorithm: Starting with the heaviest weight and working down to the lightest, put each weight into the pile that tends at each step of the way to keep the weights of the piles as equal as possible. So, in our simple five-weight example, you'd start by putting the heaviest weight, a 3-pounder, in one pile:

3

Then, to even things out, you'd place the next heaviest weight, the other 3-pounder, in the other pile:

3                    3

Then you'd take the next heaviest weight, a 2-pounder, and add it to either pile (since the piles are equal, it obviously doesn't matter which):

3, 2                    3

Then, to keep things even, you'd add a 2-pounder to the lighter pile:

3, 2                    3, 2

Finally, you'd place the remaining 2-pounder in either pile:

3, 2                    3, 2, 2

Now, this algorithm didn't produce the best solution of 3, 3 and 2, 2, 2. Nor did it produce such patently lopsided divisions as 3, 2, 2, 2 and 3. The heaviest pile the algorithm yielded was 1 pound heavier than the optimally heaviest pile of 6 pounds. Thus the algorithm was off by a factor of 1/6 (about 16 percent). Now—and here's the kicker—Graham was able to prove that for two piles and any distribution of weights whatsoever, this easy-to-apply algorithm can never be worse off than 16 percent. One can imagine trucking magnates happy with the knowledge that if their workers packed up two trucks according to Graham's simple instructions, they could never be off by more than 16 percent from the most even weight distribution— and that even if they brought in smarter workers, they could do no better.

Similar problems come up all the time in industry. Many are forms of what is known as the bin-packing problem. The idea is to pack a set of items into a set of bins, such that the total weight (or total volume) doesn't exceed some number. The problem, notes Graham, occurs in a variety of guises: "A plumber must cut a set of pipes of different lengths from a minimum number of standard-length pipes; a television network would like to schedule its commercials of varying lengths in a minimum number of program breaks; a paper manufacturer must furnish its customers with rolls of paper of different widths that he slices from a minimum number of standard rolls." Graham adds another commonplace example: "a person confronting a standard postage stamp machine with an armful of assorted letters in one hand and a handful of quarters in the other."

The lazy way to tackle the bin-packing problem is to take the items in whatever order they come and put each one into the first bin it fits in. This first-fit packing algorithm, as it is aptly called, doesn't turn out to be very effective. In 1973, Jeffrey Ullman at Princeton University proved that this algorithm can yield a solution that is off by as much as 70 percent. The algorithm performs the worst when the larger weights happen to be packed last, because new bins will need to be started before earlier bins can be filled up.

Experienced suitcase packers know that if the smaller

items are saved for the end, they can be placed in the odd gaps in nearly filled luggage. This suggests a smarter strategy: order the items from heaviest to lightest, before putting each item into the first bin that can accommodate it. Sure enough, this first-fit, heaviest-to-lightest algorithm has been shown never to be off by more than 22 percent. In 1973, David Johnson, a colleague of Graham's at AT&T, succeeded in proving that this strategy cannot be beaten in practice, that no efficient bin-packing strategy can ever be guaranteed to do better than 22 percent. That Johnson's proof took more than seventy-five pages is testimony to how thorny worst-case mathematics can be.

Counterintuitive anomalies abound in bin packing, thwarting any straightforward approach. Graham gives an example with thirty-three weights:

| 442 | 252 | 127 | 106 | 37 | 10 | 10 |
|-----|-----|-----|-----|----|----|----|
| 252 | 252 | 127 | 106 | 37 | 10 | 9 |
| 252 | 252 | 127 | 85 | 12 | 10 | 9 |
| 252 | 127 | 106 | 84 | 12 | 10 | |
| 252 | 127 | 106 | 46 | 12 | 10 | |

and bins that have a capacity of 524 pounds. The first-fit, heaviest-to-lightest strategy requires seven bins, each filled to the 524-pound capacity:

| Bin #1 | 442 | 46 | 12 | 12 | 12 | |
|--------|-----|-----|-----|-----|----|----|
| Bin #2 | 252 | 252 | 10 | 10 | | |
| Bin #3 | 252 | 252 | 10 | 10 | | |
| Bin #4 | 252 | 252 | 10 | 10 | | |
| Bin #5 | 252 | 127 | 127 | 9 | 9 | |
| Bin #6 | 127 | 127 | 127 | 106 | 37 | |
| Bin #7 | 106 | 106 | 106 | 85 | 84 | 37 |

But throw out the 46-pound weight, and the same algorithm demands not seven bins but eight!

| Bin #1 | 442 | 37 | 37 | | | | | | (unused capacity: 8) |
| Bin #2 | 252 | 252 | 12 | | | | | | (unused capacity: 8) |
| Bin #3 | 252 | 252 | 12 | | | | | | (unused capacity: 8) |
| Bin #4 | 252 | 252 | 12 | | | | | | (unused capacity: 8) |
| Bin #5 | 252 | 127 | 127 | 10 | | | | | (unused capacity: 8) |
| Bin #6 | 127 | 127 | 127 | 106 | 10 | 10 | 10 | | (unused capacity: 7) |
| Bin #7 | 106 | 106 | 106 | 85 | 84 | 10 | 10 | 9 | (unused capacity: 8) |
| Bin #8 | 9 | | | | | | | | (unused capacity: 515) |

To appreciate the paradox, imagine you're Imelda Marcos packing shoes for a weekend away from the palace. You pack the shoes from the largest to the smallest, from the thigh-high bejeweled boots to the skimpy beach thongs, as the algorithm suggests, and find that seven suitcases are required. On further reflection, you realize you don't need to take ski boots—you'll be nowhere near snow—so you dump everything out of the suitcases, put the boots aside, and carefully repack the rest. Holy Manila, you find that with one less item you need an additional suitcase, and the shoes, rather than being packed snugly, are rattling around. You curse the opposition forces that have put a hex on your packing!

Counterintuitive anomalies also arise in two-dimensional versions of bin packing, such as the cookie-cutter problem (maximize the number of cookies you can cut from a given expanse of dough) and the tiling problem (maximize the amount of floor space you can cover with a tile of fixed shape and dimensions). The absence of the third dimension doesn't make these problems any easier. When the cookies and tiles are irregularly shaped, there is no uniform strategy for maximizing the results; the gingerman problem, say, has not been addressed in the mathematical literature. But even the most regular of shapes holds surprises.

Consider the problem of tiling a huge surface with an unlimited number of one-inch by one-inch squares. If the surface is perfectly square, with sides that measure an integral number of inches, say 1,000,000, the solution is obviously a square grid of tiles, 1 million tiles long by 1 million tiles wide. Because these 1 trillion squares cover the entire surface, it is certainly the optimal tiling pattern. But say the surface, though still a perfect square, turns out to measure 1,000,000.1 inches on each side. Is it possible to cover more area than the square array of 1 trillion tiles, which leaves uncovered a thin, 1/10-inch strip along two sides? It seems unlikely that even a single tile could be added by jostling the trillion tiles so that they're no longer flush. But in 1975, to the surprise and delight of the combinatorics community, Erdős and Graham succeeded in adding not just one tile but more than 100,000 tiles, by cleverly pitching the initial 1 trillion at skewed angles with little gaps between them. Although Graham suspects that the 1,000,000,100,000-tile solution is close to optimal, Erdős and he were not able to prove that this was the best possible way to tile a large square with little squares. To this day, though, more than two decades later, no one has been able to improve on their method of square packing. *A problem worthy of attack / Proves its worth by fighting back.*

■

In late January 1987, at Graham's house, Erdős was taxing his hosts. "His toenails were bothering him," said Graham.

"They were really grungy. He wanted Fan to cut them, but that's where she drew the line." Graham put him on a plane to San Antonio to attend the annual conferences of the American Mathematical Society (AMS) and the Mathematical Association of America. San Antonio's mayor, Henry Cisneros, then a rising star in Democratic politics, had proclaimed Math Day in honor of the 2,575 mathematicians who had descended on the city. Cisneros's gesture did not advance his cause with the conferees I was with, who wondered whether he had ever met a mathematician, let alone heard of Paul Erdős. The schedule included meetings to discuss whether mathematicians should accept "Star Wars" money and whether the National Security Agency, whose code-cracking needs make it the largest employer of mathematicians in the United States, qualified for corporate membership in the AMS. But except for a few zealots, most of the mathematicians came to San Antonio not to discuss ethics and politics but to do mathematics. At physics conferences or psychoanalytic meetings, the participants do not perform experiments on subatomic particles or practice psychotherapy; they just talk about it. At mathematics conferences, the attendees actually *do* mathematics—on blackboards, napkins, placemats, and toilet-stall walls, and in their minds.

Erdős rarely attended the scheduled talks at these meetings, preferring to work simultaneously with several mathematicians in a hotel room. One of the days in San Antonio, Erdős had taken over someone else's room at the Marriott and was working on six problems with six different mathematicians, who were sprawled across two double beds and the floor. "What about 647? Is it a prime?" asked a man who looks like a plump Moses. "I can no longer do them in my head." A woman in a multicolored dress came to his rescue by pulling out a 276-page printout of all the primes up to 2 million—148,933 of them, ranging from 2 to 1,999,993. Sure enough, 647 is on the list.

Erdős didn't seem to be paying attention. He was slumped over in a chair, his head in his hands, like an invalid in a nursing home. But every few minutes he perked up and suggested a line of attack to one of his colleagues, who then scrambled to implement the master's suggestion. The others waited patiently for Erdős to have a flash of insight about their problem. Sometimes when Erdős raised his head, he fooled them. They leaned forward in anticipation of a tip. But instead of sharing a mathematical inspiration, he uttered an aphoristic statement having to do with death—"Soon I will be cured of the incurable disease of life," or, "This meeting, like life, will soon come to an end, but the meeting was much more pleasant"—and then bowed his head again. No one picked up on these comments, and the cycle of mathematical insights and reflections on death continued all morning.

"In ten years," said the man who looks like Moses, "I want you to talk to the SF on my behalf."

"What do you want from the SF?" Erdős asked.

"I want to see the Book."

"No one ever sees the Book. At most, you get glimpses."

Moses turned on the TV. "Television," Erdős said, "is something the Russians invented to destroy American education." The news came on, and Ronald Reagan filled the screen. "Eisenhower was an enthusiastic but not very good golfer," Erdős said. "Someone said at the time that it was okay to elect a golfer, but why not a good golfer? I say, it's okay to elect an actor, but why not a good actor, like Chaplin?" Reagan dissolved, and the newscast switched to a story about AIDS. "Both bosses and slaves tell me people are less promiscuous," Erdős said, "but I wouldn't know." When the conversation strayed from mathematics and death, it was a sure sign that Erdős was bored and ready to find new mathematical soulmates.

Two hours and 5 milligrams of Benzedrine later, Erdős was on a flight to Newark, back to Graham's house. From there he was going in quick succession to Memphis, Boca Raton, San Juan, Gainesville, Haifa, Tel Aviv, Montreal, Boston, Madison, De Kalb, Chicago, Champaign-Urbana, Philadelphia, and back to Newark. But Graham called Erdős's attention to a small problem in his schedule: "Two mathematicians in different states, Paul, want you to open your brain to them at the same time."

"You have heard about my mother's theorem?" Erdős said. "My mother said, 'Even you, Paul, can be in only one place at one time.' Maybe soon I will be relieved of this disadvantage. Maybe, once I've left, I'll be able to be in many places at the same time. Maybe then I'll be able to collaborate with Archimedes and Euclid."

# 4

## MARGINAL REVENGE

THOMASINA: If *you* do not teach me the true meaning of things, who will?

SEPTIMUS: Ah. Yes, I am ashamed. Carnal embrace is sexual congress, which is the insertion of the male genital organ into the female genital organ for purposes of procreation and pleasure. Fermat's last theorem, by contrast, asserts that when $x$, $y$, and $z$ are whole numbers each raised to power of $n$, the sum of the first two can never equal the third when $n$ is greater than 2. (*Pause.*)

THOMASINA: Eurghhh!

SEPTIMUS: Nevertheless, that is the theorem.

THOMASINA: It is disgusting and incomprehensible. Now when I am grown to practise it myself I shall never do so without thinking of you.

—Tom Stoppard, *Arcadia*, Act I, Scene 1

On the warm and cloudy evening of April 8, 1974, Dodger lefthander Al Downing threw a high fastball into Hank Aaron's ample strike zone. At 9:07 P.M. Aaron whipped his bat and sent the ball arcing deep into left centerfield all the way over the fence. The Atlanta hometown crowd of 53,775 roared as Aaron's 715th home run eclipsed Babe

Ruth's 1935 record of 714. "I just thank God it's all over," said Aaron, expressing the same kind of relief a mathematician might feel after doggedly solving a forty-year-old problem. The numbers 714 and 715 had been on the lips of Atlanta sports fans for months, recalled University of Georgia mathematician Carl Pomerance, and "questions like 'When do you think he'll get 715?' were perfectly understood, even with no mention made of Ruth, Aaron, or home run." Pomerance was a young assistant professor then and he noticed that the product of 714 and 715 was also the product of the first seven primes:

$$714 \times 715 = 2 \times 3 \times 5 \times 7 \times 11 \times 13 \times 17$$

A student of one of Pomerance's colleagues also found another interesting property of 714 and 715: the sum of the prime factors of 714 equaled the sum of the prime factors of 715. In other words,

$$714 = 2 \times 3 \times 7 \times 17$$
$$715 = 5 \times 11 \times 13$$
$$2 + 3 + 7 + 17 = 5 + 11 + 13$$

Pomerance called pairs of consecutive integers that had this property *Ruth-Aaron pairs*. Such pairs were exceedingly rare. Pomerance ran a computer search on the numbers below 20,000 and came up with only twenty-six Ruth-Aaron pairs, ranging from 5 and 6 to 18,490 and 18,491. Although the pairs thinned out as they became larger, as the primes themselves did, Pomerance suspected there were infinitely many Ruth-Aaron pairs, but had no idea how to prove his hunch. Pomerance wrote up his discovery in a lighthearted paper in the *Journal of Recreational Mathematics*. Within a week of publication, he received a call from Erdős, whom he had never met. The master number theorist said he had proved Pomerance's conjecture and wanted an invitation to Atlanta to demonstrate the proof. Their subsequent meeting began a collaboration that resulted in twenty-one papers.

In 1995, Erdős and Aaron were awarded honorary degrees at Emory University. Like the other honorees, Erdős wore a cap and gown, but he also wore his sandals and sat on the podium with his head in his hands, doodling during the ceremony in one of his mathematical notebooks. Pomerance told the home-run king all about Ruth-Aaron numbers. "He was a gentle man," said Pomerance, "and listened patiently as I explained that what he did also changed the life of one small mathematician." Pomerance convinced Erdős and Aaron to autograph a baseball for him. "And thus," said Pomerance, "Hank Aaron has Erdős number one."

Erdős and Pomerance did important joint work on prime numbers, building on the results of Pierre de Fermat. In a letter in 1640, Fermat posed a conjecture without offering proof: If $n$ is a prime number, then for every integer $a$, the number $a^n - a$ is a multiple of $n$. For example, because 3 is a prime, the number $a^3 - a$ should be a multiple of 3 for every $a$ whatsoever. When $a$ is 2, for instance, $2^3 - 2$ is 6. So far so good, because 6 obviously is a multiple of 3. When $a$ is 3, $3^3 - 3$ is 24, which is also a multiple of 3. When $a$ is 4, $4^3 - 4$ is 60, which again is a multiple of 3. You get the idea: you can keep doing this ad infinitum.

Fermat's statement about primes came to be called Fermat's Little Theorem, to distinguish it from his notorious Last Theorem, but its implications are by no means little. Proved by Euler nearly a century later, Fermat's Little Theorem is the basis of most contemporary tests to determine whether or not a number is prime.

To test whether $n$ is prime, see if $a^n - a$ is a multiple of $n$ for various $a$'s starting with 2. If the Fermat test fails at any point, you know the number is composite. For example, to test whether 9 is prime (and if you've made it this far in this book, you shouldn't have any doubt), see if $2^9 - 2$ is a multiple of 9. Sure enough, $2^9 - 2$ is 510, which most certainly isn't a multiple of 9, so as expected, 9 fails the primality test. If a number fails the test, it definitely isn't prime. But if it passes the test, it still might turn out not to be prime.

Twenty-five hundred years ago, Chinese number theorists mistakenly believed that a number $n$ is prime if the expression $2^n - 2$ is a multiple of $n$. So 5 is prime because $2^5 - 2$ is 30, which is a multiple of 5. And 7 is prime because $2^7 - 2$ is 126, which is a multiple of 7. While no composite number below 340 slips through the Chinese test as prime, the composite number 341 (11 times 31) does evenly divide $2^{341} - 2$. Is 341 the only prime impostor? Definitely not. There are an infinite number of odd *pseudoprimes*, and in 1950 Derrick H. Lehmer identified the first even pseudoprime: 161,038. By 1951, an infinity of even pseudoprimes were known to exist. But pseudoprimes can be weeded out by running them through the Fermat test for other values of $a$. For example, the trickster 341 fails to divide $3^{341} - 3$, and so its true composite nature is exposed.

Is there a composite number $n$ that divides not only $2^n - 2$ but also $3^n - 3$, $4^n - 4$, $5^n - 5$, $6^n - 6$, and so on? In other words, an absolute pseudoprime that fools the Fermat test for every value of $a$? The answer is yes. In 1910, Robert Carmichael, an American mathematician, found the first and smallest absolute pseudoprime: 561. For every value of $a$, Fermat's expression $a^{561} - a$ is a multiple of 561. Called *Carmichael numbers*, these absolute pseudoprimes—such as 1,105, 1,729, 2,465, 2,821, 10,585, and 15,841—prevent the

Fermat test from establishing primality beyond a shadow of a doubt. For some time it was hoped that the absolute pseudoprimes were finite and few enough in number to be enumerated in a list. Then if a candidate prime $n$ was not on the list and passed the Fermat test for successively larger values of $a$, you could be increasingly confident that $n$ was truly prime. In 1956, Erdős sketched a technique for constructing large absolute pseudoprimes and suggested, but did not prove, that such impostors come up infinitely often. In 1992, Carl Pomerance and two colleagues at the University of Georgia at long last turned Erdős's suggestion into an ironclad proof.

■

Though only a century passed before Euler proved Fermat's Little Theorem, his even simpler-sounding Last Theorem would defeat generation after generation of the best minds in mathematics for more then 350 years. In the late 1630s, Fermat claimed that he had a proof, but tantalizingly kept it to himself. The truth of the theorem was not established until 1994, when Andrew Wiles at Princeton University emerged triumphant from eight years of largely clandestine work to articulate a long and difficult proof. Erdős did not approve of Wiles's working in isolation. The theorem might have been proved earlier, Erdős felt, if Wiles had let the mathematics community in on his work. Wiles apparently feared that if he hadn't kept his work secret, others might have beaten him to the finish line or mocked him for tilting at windmills like Don Quixote. To throw his colleagues off track all those years, Wiles released a series of minor papers. "This is probably the only case I know where someone worked for such a long time without divulging what he was doing, without talking about the progress he was making," said Ken Ribet, whose own work was integral to Wiles's proof. "Mathematicians are always in communication. When you talk to other people you get a pat on the back; people tell you what you've done is important, they give you ideas. It's sort of nourishing and if you cut yourself off from this, then you are doing something that is probably psychologically very odd."

Odd or not, Wiles got the glory for slaying the most romantic unsolved problem in number theory. Never has a mathematician received more publicity. *People* magazine put him on its list of "The 25 most intriguing people of the year," the Gap offered to pay him to model jeans, and Barbara Walters invited him on her show. "Who's Barbara Walters?" asked Wiles, who had gone through life without a television.

The proof was also the subject of fake news stories that flooded the e-mail boxes of mathematicians.

Chicago, July 30. "Math hooligans are the worst," said a Chicago Police Department spokesman. "But the city learned from the Bierberbach riots. We were ready for them this time."

When word hit Wednesday that Fermat's Last Theorem had fallen, a massive show of force from law enforcement at universities all around the country headed off a repeat of the festive looting sprees that have become the traditional accompaniment to triumphant breakthroughs in higher mathematics.

Mounted police throughout Hyde Park kept crowds of delirious wizards at the University of Chicago from tipping over cars on the midway as they first did in 1976 when Wolfgang Haken and Kenneth Appel cracked the long-vexing Four Color Problem. Incidents of textbook throwing and citizens being pulled from their cars were described by the university's math department as isolated.

Pierre de Fermat was born in the South of France, in 1601, the son of a wealthy leather merchant. A smart kid who didn't demonstrate any particular affinity for numbers, Fermat was encouraged by his family to join the French civil service, through whose ranks he rose to become a judge in the service of Louis XIV. By day, Fermat condemned miscreants to be burned at the stake. At night, he kept largely to himself because French law discouraged him from fraternizing with folks who might later come before his court. The hermit nightlife suited him well. He pored over ancient history and science texts and discovered the world of mathematics. Numbers soon proved to be his chief nocturnal love. Even with Fermat's considerable mathematical skills, it would have been difficult for him to make a living as a mathematician. In the early seventeenth century, when Europe was emerging from the Dark Ages, mathematics was not thought of as a noble profession. Those who were mathematically inclined might find employment as accountants, working in secret on the financial affairs of wealthy merchants. The tradition of clandestine work spread even to mathematicians who weren't cooking the books of rich people, and Fermat was no exception. He tended to keep his mathematical discoveries to himself, although he did occasionally like to provoke his colleagues by boasting in correspondence that he had proved thus-and-such a theorem, withhold the details, and challenge them to find the proof.

In 1637 he was reading a Latin translation of *Arithmetica*, a treatise on number theory penned by a mysterious algebraist named Diophantus. Little is known about this ancient Greek thinker except that he lived in Alexandria, the leading hub of mathematical activity for seven hundred years (from the days of Euclid to the death of Hypatia, the first known female mathematician, early in the fifth century). Exactly when Diophantus resided in Alexandria is unclear; historians of science usually peg his residence to the middle of the third century but concede that the evidence is so slight that he could have lived in Alexandria any time over a five-hundred-year period.

Diophantus's *Arithmetica* originally consisted of thirteen books, only six of which have made it to modern times. Some of the volumes may have been lost to the two conflagrations that consumed the Library of Alexandria: In 47 B.C., Julius Caesar torched Cleopatra's fleet and the flames leaped to the nearby library; and in the year 642, Muslim invaders who ransacked the city burned many of the remaining books. But the six volumes of *Arithmetica* that managed to survive are gems. They show that Diophantus was even more conservative than Pythagoras; while Pythagoras dismissed irrational numbers like the square root of 2 but cheerfully admitted fractions to the brotherhood of numbers, Diophantus favored whole numbers. He was famous for solving problems, now called *Diophantine equations*, that had whole numbers as their solutions.

One such chestnut, preserved in the *Greek Anthology*, a problem book from the fifth or sixth century, gives a rare glimpse of Diophantus's life:

> God granted him to be a boy for the sixth part of his life, and adding a twelfth part to this, He clothed his cheeks with down. He lit him the light of wedlock after a seventh part, and five years after his marriage, He granted him a son. Alas! Late-born wretched child; after

attaining the measure of half his father's life, chill Fate took him. After consoling his grief by this science of numbers for four years he ended his life.

Aside from providing biographical information on the elusive Diophantus, the riddle is notable as one of the earliest expressions of the sentiment that mathematics can be a soothing escape from the harsh realities of life. The information in the riddle can be translated into a simple algebraic equation, a Diophantine equation, in which $D$ represents the age Diophantus lived to:

$$D = D/6 + D/12 + D/7 + 5 + D/2 + 4$$

Every school kid who paid attention in algebra class knows how to solve this equation for $D$. If the riddle is to be believed, Diophantus lived to eighty-four, a ripe old age for a man of the third century.

The riddle equation is said to be of "the first degree," meaning that the variable $D$ is not raised to any power. Diophantus himself usually focused on more complicated equations, where the variables were squared, cubed, or raised to higher powers. In *Arithmetica*, Diophantus discussed at length the Pythagorean theorem: the familiar statement that in any right-angled triangle, the square of the hypotenuse equals the sum of the squares of the other two sides. Diophantus observed that there were an infinite number of Pythagorean triplets, whole numbers $x$, $y$, and $z$ that solved the equation $x^2 + y^2 = z^2$. (Tables of numbers etched into Plimpton 232, a clay tablet dating from 1900 B.C., show that the Babylonians actually knew of these triplets—such as, 3, 4, 5 and 5, 12, 13—a thousand years before Pythagoras.)

When Fermat read Diophantus's discussion, he wondered whether or not the same equation had whole number solutions for exponents greater than 2. In other words, were there any triplets that satisfied the requirement that $x^3 + y^3 = z^3$ or $x^4 + y^4 = z^4$ or, for that matter, $x^{707} + y^{707} = z^{707}$? While there was no limit to the number of solutions when the exponent was 2, Fermat made the extraordinary claim that for all higher exponents there were absolutely no solutions at all. In the margin of his copy of *Arithmetica*, he wrote (translated from the Latin): "It is impossible for a cube to be written as a sum of two cubes or a fourth power to be written as the sum of two fourth powers or, in general, for any number which is a power greater than the second to be written as a sum of two like powers." To this far-reaching statement—Fermat's Last Theorem—he appended the comment: "I have a truly marvelous demonstration of this proposition which this margin is too narrow to contain." Fermat died in 1665, and his theorem would have gone with him to the grave were it not for his eldest son, Clement-Samuel, who dusted off his father's dog-eared copy of *Arithmetica* and published the marginalia in 1670.

The marginalia caught on, driving scores of number theorists to distraction. The names of the men and women who tried to prove Fermat's Last Theorem over the centuries read like a *Who's Who of Mathematics*. Fermat himself did sketch a proof for the case where the exponent was 4. The great Leonhard Euler fell under the problem's spell, but frustrated by lack of progress, in 1742 he sent a friend to search Fermat's old house for whatever it was that wouldn't fit in the margin. When the friend came back empty-handed, Euler buckled down and demonstrated that for the case of cubes, no solutions were possible. So $x^3 + y^3 = z^3$ and $x^4 + y^4 = z^4$ had fallen. What remained were an infinite number of other cases:

$$x^5 + y^5 = z^5$$
$$x^6 + y^6 = z^6$$
$$x^7 + y^7 = z^7$$
$$x^8 + y^8 = z^8$$

and so on.

The next real progress was made by Sophie Germain, who was born in Paris in 1776, the daughter of a wealthy banker. When the French Revolution raged in the streets of Paris in 1789, Germain was confined to the safety of her home, where she passed the days in her father's library. There the thirteen-year-old was drawn to mathematics by a romantic account she read of Archimedes' death. The man from Syracuse was said to be sitting on the ground, contemplating geometric figures sketched in the dirt, unaware of a Roman assault on his city. A soldier came up from behind him and defiantly stomped on his figures. "Don't step on my circles!" Archimedes protested, whereupon the soldier drew his sword and executed the seventy-five-year-old mathematician. Germain wanted to devote herself to a subject that could be so engrossing it would distract from dangers to life and limb, like the Reign of Terror just outside her house.

Germain's parents, however, were not encouraging. They hid the math books because they shared the middle-class view of the time that ladies should not bother themselves with numbers. Germain found the books and pored over them, away from disapproving eyes, in the middle of the night. When her parents caught on, they took extreme steps to thwart her. Once she was in bed, they evidently made it harder for her to get out from under the covers by taking away her clothes as well as the sources of heat and light. But when she defeated them again, with a secret stash of candles, they came to tolerate her passion for math and eventually even support it.

French society was slower to become enlightened. Women were banned from the École Polytechnique, which opened in 1794, but the resourceful eighteen-year-old managed to get hold of the lecture notes. Passing herself off as "Antoine-August Le Blanc," a promising student who had dropped out of the École Polytechnique, she started corresponding with leading mathematicians. In her twenties, Germain came up with a general technique for proving a special case of Fermat's Last Theorem in which the exponent $n$ was a certain kind of prime number. (In the mathematical literature, such primes—which have the property that if you double them and add 1, you get another prime—came to be called *Sophie Germain primes*. The first Sophie Germain prime is 2, because 2 doubled plus 1 is 5, which is prime. The largest-known Sophie Germain prime, discovered on January 19, 1998, weighs in at 5,122 digits. It is equal to $92,305 \times 2^{16,998} + 1$, which means that doubling $92,305 \times 2^{16,998} + 1$ and adding 1 yields another prime: $92,305 \times 2^{16,999} + 3$. Whether the supply of Sophie Germain primes is inexhaustible, like the primes themselves, is not yet known.)

Under her masculine nom de plume, Germain shared her work on Fermat's Last Theorem in a letter to University of Göttingen professor Carl Gauss, one of the most famous mathematicians and astronomers of the day. The letter began with an apology: "Unfortunately, the depth of my intellect does not equal the voracity of my appetite, and I feel a kind of temerity in troubling a man of genius." Gauss wrote back with words of encouragement: "I am delighted that arithmetic has found in you so able a friend."

Gauss did not learn of her true identity until 1807. When Napoleon's forces moved into Prussia, Germain feared that Gauss might come to the same end as Archimedes, so she asked a commander she knew in the French Army to ensure his safety. He sought out Gauss and told him that his life was safe on account of the intercession of one Sophie Germain. Gauss had no idea who his mysterious benefactor was until Germain admitted her deception in a subsequent letter. Gauss was delighted by the turn of events: "A taste for the abstract sciences in general and above all the mysteries of numbers is excessively rare.... But when a person of the sex which, according to our customs and prejudices, must encounter infinitely more difficulties than men to familiarize herself with these thorny researches, succeeds nevertheless in surmounting these obstacles and penetrating the most obscure parts of them, then without doubt she must have the noblest courage, quite extraordinary talents and superior genius."

For all their corresponding and mutual admiration, Gauss and Germain never met. He persuaded the University of Göttingen to grant her an honorary degree, but before she could make the trip from France, she died at the age of fifty-five after a two-year battle with breast cancer. "Being nice to her was not typical behavior for Gauss," Erdős said. "Gauss was mean, although not as mean as Newton. Often when students shared their work with Gauss, he would tell them he had done it all before. Maybe he had. Maybe he hadn't. But it was wrong of him to squash the youthful enthusiasm of students."

In the nineteenth century, after Germain's pioneering work, the next significant progress on Fermat's Last Theorem was made in the late 1840s by the German number theorist Ernst Eduard Kummer, who taught high school mathematics and then cannonball ballistics at Berlin's war college. Though many number theorists flaunted the lack of applications of their work, Kummer was proud to use his first-rate mathematical mind in the war service of his country. He wanted revenge. At the age of three, he had seen his home town of Sorau overrun by Napoleon's forces. A typhoid epidemic introduced by the French troops added to the carnage; Kummer's father, the town physician, struggled mightily to contain the disease, but succumbed to it himself.

When Kummer was not busy keeping the French military in its place, he was busy doing the same to French mathematicians. The French Academy of Sciences had offered 3,000 francs for a proof of Fermat's Last Theorem, and Parisian café society was buzzing about a nasty race to complete the proof between two local mathematicians, Gabriel Lamé and Augustin-Louis Cauchy. Word of their respective methods spread to Germany, and Kummer, realizing that both their approaches were fatally flawed, fired off a letter to the French Academy. Lamé and Cauchy were suitably embarrassed, and turned their mathematical attention to more tractable problems.

Kummer was able to demonstrate that Fermat's Last Theorem was true for an infinite number of cases, in which the exponent was a particular type of prime number, which he called a "regular prime." Kummer's method was the first general attack on Fermat's Last Theorem, although its generality was limited by the nagging presence of so-called irregular primes, which were not really all that irregular. From 1 to 100, for instance, three primes are irregular: 37, 59, and 67. (Never mind what regular primes are; the definition is too complex and would defeat even the great mathematical expositor Joseph-Louis Lagrange, who "believed that a mathematician has not thoroughly understood his own work till he has made it so clear that he can go out and explain it effectively to the first man he meets on the street." By that wishful criterion, most mathematicians would have to close up shop.)

For all of Kummer's work on the cutting edge of number theory, he was apparently rather bad at elementary arithmetic. One story has him standing before a blackboard, trying to compute 7 times 9. "Ah," Kummer said to his high school class, "7 times 9 is eh, uh, is uh. . . ." "61," one of his students volunteered. "Good," said Kummer, and wrote 61 on the board. "No," said another student, "it's 69." "Come, come, gentlemen," said Kummer, "it can't be both. It must be one or the other." (Erdős liked to tell another version of how Kummer computed 7 times 9: "Kummer said to himself, 'Hmmm, the product can't be 61 because 61 is a prime, it can't be 65 because that's a multiple of 5, 67 is a prime, 69 is too big—that leaves only 63.' ")

The twentieth century saw a flurry of activity, much of it misguided, on Fermat's Last Theorem. In 1908, the German Academy of Sciences outdid the French Academy by offering a prize of 100,000 marks. The German prize was funded by Paul Wolfskehl, a Darmstadt industrialist and amateur mathematician who claimed to owe his life to number theory as much as Archimedes owed his death to geometry. Spurned by the woman of his dreams, the temperamental Wolfskehl plunged into such a state of despondency that only suicide seemed to promise relief. A compulsive fellow, Wolfskehl couldn't just kill himself on the spot but had first to get his affairs in order and set an exact time for blowing his brains out. His affairs took less time to straighten out than he had allowed, and with a few hours to kill before the execution hour, Wolfskehl went into his library, thumbed through his math books, and stumbled on Fermat's Last Theorem. He got so caught up trying to prove it that he missed his date with death. The contemplation of difficult mathematics, Wolfskehl realized, was far more rewarding than the love of a difficult woman, so he resumed life with relish and bankrolled the prize for putting Fermat's Last Theorem to rest.

But the theorem wouldn't rest. The 100,000-mark Wolfskehl Prize spurred amateur and professional mathematicians alike to hunt for a proof—or, conceivably, a counterexample. In 1908, the first year of the prize, 621 alleged proofs were submitted. In the 1970s, the submissions were still coming in strong. Nonsensical manuscripts were sent back immediately, while ones that looked like mathematics were examined by someone in the University of Göttingen math department, and "at the moment," wrote F. Schlichting in 1974, "I am the victim. There are about 3 to 4 letters to answer per month, and there is a lot of funny and curious material arriving, e.g., like the one sending the first half of his solution and promising the second if we would pay 1000 DM in advance; or another one, who promised me ten percent of his profits from publications, radio and TV interviews after he got famous, if only I would support him now; if not, he threatened to send it to a Russian mathematics department to deprive us of the glory of discovering him. From time to time someone appears in Göttingen and insists on personal discussion." At other universities, crackpots often were brushed off with a form response: "I have a marvelous demonstration of the incorrectness of your attempted proof which this page is not large enough to contain." But mainstream mathematicians continued to demonstrate that Fermat's Last Theorem was true for individual exponents. By 1993 it was known that if a counterexample existed, the exponent would have to exceed 4 million.

■

The man whose destiny it was to slay Fermat, Andrew Wiles, was the son of an Oxford theologian. He was drawn to mathematics by reading a romantic account of Fermat, just as Sophie Germain before him had been drawn to the subject by reading a romantic account of Archimedes. At the age of ten, Wiles read Eric Temple Bell's book, *The Last Problem* (1961). Bell, a leading figure in American mathematics for the first half of this century, had a rare gift for words as well as numbers. Those who have witnessed the deep truths of mathematics, Bell wrote, "have experienced something no jellyfish has ever felt." He had a knack for pithily summing up a man's character: Pythagoras, Bell said, whose mysticism had hobbled his mathematics, was "one-tenth genius, nine-tenths sheer fudge." And if Bell's prose was at times flowery, *The Last Problem* and his better-known 1937 work, *Men of Mathematics*, sowed the seeds of mathematical interest in three generations of readers.

Bell's innovation in 1937, which doesn't sound like much in our current tabloid age but was novel in its time, was to write jauntily not just about the work of mathematicians but also about their lives and loves. Take his description of René Descartes: "In spite of his lofty thoughts he was no gray-bearded savant in a dirty smock, but a dapper well-dressed man of the world, clad in fashionable taffeta and sporting a sword.... To put the finishing touches on his elegance he crowned himself with a sweeping, broad-brimmed, ostrich-plumed hat. Thus equipped he was ready for the cutthroats infesting church, state, and street. Once when a drunken lout insulted Descartes' lady of the evening, the irate philosopher went after the rash fool...and having flicked the sot's sword out of his hand, spared his life, not because he was a rotten swordsman, but because he was too filthy to be butchered before a beautiful lady." On reading Bell, only the braindead would fail to be drawn to mathematics, a world inhabited by sword-wielding swells.

When the young Wiles read Bell's account of Fermat's Last Theorem, he was hooked. The theorem, recalled

Wiles, "looked so simple, and yet all the great mathematicians in history couldn't solve it. Here was a problem that I, a ten-year-old, could understand and I knew from that moment that I would never let it go. I had to solve it." Bell's account of Fermat came out posthumously, in 1961, and offered the prediction that human civilization would extinguish itself in nuclear war before getting around to solving Fermat's Last Theorem. If Bell had lived a few decades longer, wondered Berkeley mathematics professor Ken Ribet, "it is an interesting question whether he would have been more surprised at humanity's continuing survival or at the announcement on June 23, 1993, of a proof of Fermat's Last Theorem." Wiles had worked on the proof as a teenager, but making no progress, turned to a hot research topic in contemporary mathematics—the study of what are called *elliptic curves* (which are objects more complex than a simple ellipse)—so that he could complete his dissertation and earn his Ph.D. at Cambridge. Wiles put Fermat on the back burner for the next decade.

In 1986 Ken Ribet surprised himself by proving that a certain unproven conjecture, the so-called Taniyama-Shimura conjecture, implied Fermat's Last Theorem. In other words, the problem of proving Fermat's Last Theorem was now reduced to the problem of proving the Taniyama-Shimura conjecture, which itself was no small task because it had been an open problem for thirty years. And those who wanted to prove the Taniyama-Shimura conjecture could no longer rely on the insights of one of the two men who first articulated it. The allure of mathematics could not rescue Yutaka Taniyama from the depths of mental anguish as it had rescued Paul Wolfskehl. Taniyama, one of the most brilliant mathematical minds of postwar Japan, killed himself in 1958, at the age of thirty-one.

When Wiles learned of Ribet's result, the flames of

Fermat consumed him anew, but this time he had a powerful new weapon in his arsenal: The Taniyama-Shimura
conjecture involved elliptic curves and he knew all about
elliptic curves from his doctoral work. Still, Wiles wasn't
sure he'd succeed. "Of course the Taniyama-Shimura conjecture had been open for many years," Wiles said later.
"No one had had any idea how to approach it but at least
it was mainstream mathematics. I could try and prove results, which, even if they didn't get the whole thing, would
be worthwhile mathematics. I didn't feel I'd be wasting my
time. So the romance of Fermat which had held me all my
life was now combined with a problem that was professionally acceptable." He holed up in his attic office for what
would be seven years of secret work, and read everything
he could find relating to the Taniyama-Shimura conjecture.
"For the first few years I knew I had no competition," he
said. "I knew I had no competition, since I knew that nobody—me included—had any idea where to start." Gradually, though, the pieces started falling into place. Wiles
likened the experience to "entering a darkened mansion.
You enter a room, and you stumble months, even years,
bumping into the furniture. Slowly you learn where all the
pieces of furniture are, and you're looking for the light
switch. You turn it on and the whole room is illuminated.
Then you go onto the next room and repeat the process."

By 1993, the entire mansion was fully illuminated. Still
he kept most everyone else away from the mansion, as he
checked and rechecked his 200-page proof. In June, he returned to his alma mater Cambridge to deliver a series of
lectures with the slyly unassuming title "Modular Forms,
Elliptic Curves, and Galois Representations." For three days
he delivered his talks without explicitly saying what they
were leading up to. By the last day the assembled crowd
of mathematicians had swelled to standing room, as many

anticipated where he was heading. "There was only one possible climax, only one possible end to Wiles's presentation," said Ribet. "I came relatively early and I sat in the front row. . . . I had my camera with me just to record the event. There was a very charged atmosphere and people were very excited. We certainly had the sense that we were participating in an historical moment. People had grins on their faces before and after the lecture. The tension had built up over the course of several days." As the end of the last lecture hour approached, Wiles wrote one last statement on the blackboard and said, softly, "This proves Fermat's Last Theorem. I think I'll stop here." For a moment the lecture hall was eerily silent. Then thunderous applause broke out, and Wiles was treated to a standing ovation.

But his emotions were mixed. He felt a kind of sadness. "All number theorists, deep down, feel that," he said. "For many of us, his problem drew us in and we always considered it something you dream about but never actually do. There is a sense of loss, actually."

On June 24, Wiles made the front page of the *New York Times* (under the headline AT LAST, SHOUT OF "EUREKA!" IN AGE-OLD MATH MYSTERY) and was hailed the world over as "the mathematical dragon-slayer." The shy, bespectacled Wiles became an overnight celebrity, which was all the more remarkable considering that, as Ribet told the *Times*, only one tenth of 1 percent of mathematicians could understand the very technical proof. "And that percentage," said Graham, "is probably an overstatement."

The proof was technical, indeed. So technical in fact that no one initially saw that the dragon wasn't actually dead yet. By late August, one of Wiles's colleagues quietly called attention to a disturbing hole in the proof. Wiles spent the fall trying to plug it, but by December, as word of the problem raged through the e-mail rumor mill, Wiles publicly acknowledged that his proof was flawed—a sad admission for a man who had been hailed the greatest mathematician of the twentieth century. This time around he did not work in isolation. He enlisted the help of a Cambridge colleague, Richard Taylor, and by September 1994—fourteen months after the original "proof"—the hole was patched. Fermat's Last Theorem was now officially proved.

In June 1997, Wiles traveled to Göttingen University, Gauss's old stomping ground, to collect the Wolfskehl Prize. If the mark had not been devalued during the hyperinflation of the 1920s, the prize would have been worth about $2 million in 1997 dollars. Still, Wiles walked away with $50,000—and the satisfaction that the dragon was really dead. "Having solved this problem, there's certainly a sense of freedom," he said. "I was so obsessed by this problem that for eight years I was thinking about it all of the time—when I woke up in the morning to when I went to sleep at night. That particular odyssey is now over. My mind is at rest."

Wiles, in disposing of Fermat's conundrum at the age of forty-one, was held up as a welcome exception to the rule that only the young do stellar mathematics. "If I've helped to counter that idea," said Wiles, "I'm very happy." Erdős was amused by talk of Wiles's age. "If he's considered old," said Erdős, "what does that make me? Extinct?" Erdős admired Wiles for slaying the dragon, but he did not pretend to understand the proof. And he was happy to learn that in eight years of working on the problem, Wiles never once used a computer.

Wiles's proof, which threw the entire kitchen sink of complex twentieth-century techniques at the problem, couldn't possibly be the proof that spilled out of Fermat's margin. The question remains whether or not Fermat really had a proof. Was he playing a mischievous trick on posterity with his marginal note or did he actually believe he had a proof that may in fact have been flawed, unbeknownst to him? "Fermat's judgment was known to be fallible," said Erdős. For example, in 1640 Fermat conjectured he had a formula $2^n + 1$ that would always yield prime numbers when the exponent $n$ is itself a power of 2. So for $n = 2$, $2^2 + 1$ is 5, which is prime. When $n = 2^2$, or 4, $2^4 + 1$ is 17, also a prime. When $n = 2^3$, or 8, $2^8 + 1$ is yet another prime, 257. And when $n = 2^4$, or 16, $2^{16} + 1$ is still another prime, 65,537. "A century passed," said Erdős, "before Euler found a counterexample." In 1732 Euler showed that the formula $2^n + 1$ failed to yield a prime in the case of $n = 2^5$, or 32. "The number $2^{32} + 1$ turns out to be equal to 641 times 6,700,417," Erdős recalled. "So maybe Fermat was mistaken in his belief that he had a proof. When I have a chance, I will ask him. When I go to the place where you don't need a passport for the return because there is no return."

But Euler was fallible, too. He noted that although Fermat forbade a cube to be the sum of two cubes, a cube could certainly be the sum of three cubes, as in

$$3^3 + 4^3 + 5^3 = 6^3$$

So Euler generalized Fermat's Last Theorem, claiming that three fourth powers can never sum to a fourth power, four fifth powers can never sum to a fifth power, and in general that $n - 1$ $n$th powers can never sum to an $n$th power, but it is always possible for $n$ $n$th powers to do so.

Euler's claim was wrong, but no one knew that until 1966, when Leon J. Lander and Thomas R. Parkin produced a counterexample for fifth powers:

$$27^5 + 84^5 + 110^5 + 133^5 = 144^5$$

In 1988, Noam Elkies at Harvard followed with a counterexample for fourth powers:

$$2{,}682{,}440^4 + 15{,}365{,}639^4 + 18{,}796{,}760^4 = 20{,}615{,}673^4$$

And soon afterward another mathematician found the smallest possible exception for fourth powers:

$$95{,}800^4 + 217{,}519^4 + 414{,}560^4 = 422{,}481^4$$

∎

Euler, though capable of making mistakes, was the greatest number theorist of the eighteenth century. And in Erdős's eyes, he deserved credit for doing mathematics up until the bitter end. On September 18, 1783, after calculating the orbit of the recently discovered planet Uranus, Euler paused to play with his grandson and sip a cup of tea. Pipe in hand, he suffered a fatal stroke, getting out the last words, "I die."

"I told this story once at a lecture," said Erdős, "and some wise guy shouted out, 'And another conjecture of Euler's is proven.' I want to leave like Euler. I want to be giving a lecture, finishing up an important proof on the blackboard, when someone in the audience shouts out, 'What about the general case?' I'll turn to the audience and smile. 'I leave that to the next generation,' I'll say, and then I'll keel over."

# 5

---

## "GOD MADE THE INTEGERS"

A graduate student at Trinity
Computed the square of infinity
But it gave him the fidgits
To put down the digits,
So he dropped math and took up divinity.

—Anonymous

The history of Fermat's Last Theorem is obviously engaging, but what is it about the theorem itself that makes it mathematically interesting? Or, for that matter, is it actually mathematically interesting? The great Gauss, for one, didn't think it was, despite his encouragement of Sophie Germain's interest in the subject. In 1816, a fellow astronomer informed the cash-starved Gauss that the French Academy of Sciences was offering 3,000 francs for a proof of the theorem, concluding his letter: "It seems right to me, dear Gauss, that you should get busy about this." Gauss turned him down, writing back: "I am very much obliged for your news concerning the Paris prize. But I confess that Fermat's theorem as an isolated proposition has very little interest for me, because I could easily lay down a multitude of such propositions, which one could neither prove nor dispose of." Gauss's haughtiness aside, his response does raise the question of what constitutes a meaningful mathematical theorem.

When Stan Ulam gave a talk at the Institute for Advanced Study on the twenty-fifth anniversary of the computer, he made a mental calculation of the number of theorems published yearly in mathematical journals and threw out the number 100,000. "The audience gasped," Ulam recalled. "The next day two of the younger mathematicians in the audience came to tell me that impressed by this enormous figure they undertook a more systematic and detailed search in the Institute library. By multiplying the number of journals by the number of yearly issues, by the number of papers per issue, and the average number of theorems per paper, their estimate came to nearly two hundred thousand theorems a year. Such an enormous number should certainly give food for thought. . . . If the number of theorems is larger than one can possibly survey, who can be trusted to judge what is 'important'?"

"Today upwards of a quarter million theorems are published a year," said Graham. "With our concept of proof, mathematicians demand a little more rigor than other scientists do. But who reads all these theorems? Proof by authority still goes a long way—that you believe a proof because you believe in the person who did the proving or the person who examined the proof. Even Erdős would say, 'I believe thus-and-such because so-and-so says it's true.' Erdős accepted the truth of the Four Color Map Theorem because someone he trusted checked the proof. Some areas of mathematics are so obscure or so new that people in these areas have no one to share their results with. I heard someone say once that an area of math isn't important until at least one hundred people are working in it. But every area has to start somewhere." The line between recreational mathematics and serious number theory is often fuzzy. No one is working today on Ruth-Aaron numbers, so they seem squarely in the recreational camp. Ruth-Aaron numbers, funny though their history may be, haven't led to significant new ideas or revealed new connections between old concepts.

*Smith numbers*, which have been cited in the mathematical literature only marginally more often than Ruth-Aaron numbers, began with a phone number. In 1982, Albert Wilansky at Lehigh University noticed (and only he knows how) that the phone number of his brother-in-law, H. Smith, had the peculiar property that the sum of its digits was equal to the sum of the digits of its prime factors. Got that? Smith's phone number, 493-7775, could be expressed as the product of prime factors: $4,937,775 = 3 \times 5 \times 5 \times 65,837$. The sum of the original digits $(4 + 9 + 3 + 7 + 7 + 7 + 5)$ was 42, as was the sum of the digits of the prime factors $(3 + 5 + 5 + 6 + 5 + 8 + 3 + 7)$. Although Wilansky's insight certainly doesn't rank with Erdős's proof of Chebyshev's theorem, it did get written up in the *Two-Year College Mathematics Journal*. Wilansky cited other Smith numbers and observed that he knew of none larger than his brother-in-law's phone number. For example, 9,985, which equals $5 \times 1,997$, is a Smith number, because $9 + 9 + 8 + 5 = 5 + 1 + 9 + 9 + 7$. And so is 6,036, because $6,036 = 2 \times 2 \times 3 \times 503$ and $6 + 0 + 3 + 6 = 2 + 2 + 3 + 5 + 0 + 3$. The smallest known Smith number is 4 because it is equal to the sum of the digits of its prime factors, 2 and 2.

Most mathematicians would agree that Smith numbers, though amusing in their origin, are mathematical dead-ends, parlor tricks disconnected from other ideas. And yet a year after the Smith's debut, Sham Oltikar and Keith Wayland of the University of Puerto Rico proved three theorems that made Smith numbers somewhat less silly by tying them into other more familiar concepts. Oltikar and Wayland found that Smith numbers could easily be generated from prime repunits (short for "repeated units"), which are primes like 11 and 1,111,111,111,111,111,111, all of whose digits are 1. These mathematicians proved that Smith numbers could be constructed by multiplying any prime repunit (greater than 11) by 3,304. And more important, they proved that every prime consisting entirely of 1's and 0's has a multiple that is a Smith number. Oltikar and Wayland asked whether there were infinitely many Smith numbers. No one then knew. A positive answer, they noted, could "be obtained by showing that there are infinitely many primes consisting only of the digits zero and one, an interesting and challenging problem in itself." Wayne McDonald soon proved that the supply of Smith numbers was inexhaustible. The proof, though, did not provide a recipe for how to construct every possible Smith number. Despite the missing recipe, mathematicians are not exactly tripping over themselves to keep up with the Smiths. Such numbers don't have anywhere near the one hundred devotees that Graham's colleague demands.

■

Insights and connections—that's what mathematicians look for. Carl Friedrich Gauss, who was born in 1777 in Braunschweig, Germany, the son of a masonry foreman, was a master at exposing unsuspected connections. Like Erdős, Gauss was a mathematical prodigy, and in his old age he liked to tell stories of his childhood triumphs. Like the time, at the age of three, he spotted an error in his father's ledger and stopped him just as he was about to overpay his laborers. Like the fact that he could calculate before he could read.

And he certainly could calculate. At the age of ten, he was a show-off in arithmetic class at St. Catherine elementary school, "a squalid relic of the Middle Ages . . . run by a virile brute, one Büttner, whose idea of teaching the hundred or so boys in his charge was to thrash them into such a state of terrified stupidity that they forgot their own names." One day, as Büttner paced the room, rattan cane in hand, he asked the boys to find the sum of all the whole numbers from 1 to 100. The student who solved the problem first was supposed to go and lay his slate on Büttner's desk; the next to solve it would lay his slate on top of the first slate; and so on. Büttner thought the problem would preoccupy the class, but after a few seconds Gauss rushed up, tossed his slate on the desk, and returned to his seat. Büttner eyed him scornfully, as Gauss sat there quietly for the next hour while his classmates completed their calculations. As Büttner turned over the slates, he saw one wrong answer after another, and his cane grew warm from constant use. Finally he came to Gauss's slate, on which was written a single number, 5,050, with no supporting arithmetic. Astonished, Büttner asked Gauss how he did it, "and when Gauss explained it to him," said Erdős, "the teacher realized that this was the most important event in his life and from then on worked with Gauss always," plying him with textbooks, for which "Gauss was grateful all his life."

What was Gauss's trick? In his mind he apparently pictured writing the summation sequence twice, forward and backward, one sequence above the other:

$$1 + 2 + 3 + 4 + \ldots + 97 + 98 + 99 + 100$$
$$100 + 99 + 98 + 97 + \ldots + 4 + 3 + 2 + 1$$

Gauss realized that you could add the numbers vertically instead of horizontally. There are 100 vertical pairs, each summing to 101. So the answer is 100 times 101 divided by 2, since each number is counted twice. Gauss easily did the arithmetic in his head.

"What makes Gauss's method of calculation so special," said Graham, "is that it doesn't just work for this specific problem but can be generalized to find the sum of the first 50 integers or the first 1,000 integers or the first 10,000 integers or whatever number you want. Gauss found a very nice way of showing that if you add all the numbers from 1 up through any number $n$, the answer is $n$ times $n$ plus 1, all divided by 2. This method of summing such a series is really straight from the Book."

Mathematics is about finding connections, between specific problems and more general results, and between one concept and another seemingly unrelated concept that really is related. No mathematical concept worth its salt stands in isolation. Not Fibonacci numbers. Not the well-known constant $\pi$. Not the prime numbers. Even Wiles's proof of Fermat's Last Theorem, difficult though it is, reveals deep connections between the algebra of Diophantine equations and the geometry of elliptic curves.

The Fibonacci series 1, 2, 3, 5, 8, 13, 21, 34, 55, 89, 144, 233 . . . arose in a problem about sexually active bunnies but has since come up again and again in design, both natural and man-made. The seeds in a sunflower, for example, are always positioned along two sets of interweaving spirals, one set turning clockwise, the other counterclockwise. The numbers of spirals in the two sets are not the same; in fact, they are always consecutive Fibonacci numbers. So, if there are 144 clockwise spirals, there are always either 89 or 233 counterclockwise spirals. The Fibonacci series has also come up in man-made design because as the series approaches infinity, the ratio between two consecutive terms approaches the "golden ratio"—the ideal proportions of a rectangle that the Greeks favored in painting and classical architecture (the Parthenon, for example). Indeed, Fibonacci numbers have so many connections to other things that an entire journal, the *Fibonacci Quarterly*, is devoted to keeping up with them.

The famous number $\pi$, the nonrepeating, nonterminating decimal that begins 3.141 . . . , sprang up in the study of circles. The Greeks recognized that in any circle whatsoever, the ratio of the circumference to the diameter has a constant value, $\pi$. But $\pi$ also comes up in all kinds of situations that have nothing to do with circles. Euler, for one, discovered that the infinite series formed by summing the reciprocals of squares was linked to $\pi$:

$$\pi^2/6 = 1/1^2 + 1/2^2 + 1/3^2 + 1/4^2 + 1/5^2 + \ldots$$

And in 1777, Comte de Buffon gave a formula involving $\pi$ for computing the probability that a needle dropped on an array of parallel lines will overlap one of the lines. (Buffon was a mathematical provocateur who stunned his contemporaries with an estimate that Earth was not 6,000 years old, as the Bible said, but 75,000 years.)

With the Prime Number Theorem, Gauss was able to connect the distribution of prime numbers to logarithms and the famous constant $e$. For those who slept through high school algebra, or didn't get that far, the logarithm of a given number is the power to which a fixed number called the *base* must be raised in order to get the given number. Thus, the base-10 logarithm (abbreviated $\log_{10}$) of 100 is 2, because $10^2 = 100$. By the same token, $\log_{10}$ of 1,000 is 3, because $10^3 = 1,000$. Note how slowly logarithms grow. From 100 to 1,000, $\log_{10}$ only goes up one integer, from 2 to 3.

As residents of California are all too aware, the Richter

scale of earthquake magnitude is logarithmic: each step on the scale, say, from 2 to 3, corresponds to a tenfold increase in quake power, while a jump from 2 to 4 corresponds to a hundredfold increase. Logarithmic growth (and *growth* hardly seems the right word because the logarithms increase ever so slowly) is the inverse of exponential growth. You can also have logarithms in other bases. For example, the base-2 logarithm of 1,000 is a number greater than 9 and just shy of 10, because $2^9 = 512$ and $2^{10} = 1,024$. The so-called *natural* logarithm of a given number (abbreviated $\log_e$) is defined as the power to which $e$ must be raised in order to get the given number.

Now what is this constant $e$? The number $e$, like $\pi$, is a nonrepeating, nonterminating decimal, which Euler calculated to twenty-three places:

$$2.71828182845904523536028\ldots$$

It is a number generated by an infinite series:

$$e = 1 + 1/1 + 1/(1 \times 2) + 1/(1 \times 2 \times 3) + 1/(1 \times 2 \times 3 \times 4) + 1/(1 \times 2 \times 3 \times 4 \times 5) + \ldots$$

The number $e$ may not seem all that "natural," but it is described as such because it comes up often in the mathematical modeling of such basic processes of life as growth and decay. Not to mention in another basic area over which most humans, with the exception of Erdős, obsess—namely, money. The number $e$ is central, for instance, to the formula for compound interest. Suppose you invest $1.00 in a bank that promises 100 percent interest compounded annually. You'll have doubled your money at the end of a year. Another bank offers 100 percent interest compounded every six months. That's a better deal, because at the end of six months you'll get interest equal to 50 percent of the investment, or 50 cents. But you'll be earning interest on the interest, so that at the end of the full year you'll have a total of $2.25. How about 100 percent interest compounded quarterly? At the end of a year you'll have $2.44. How about 100 percent interest compounded eight times annually? You'll earn $2.57 in a year. What if the Bank of Erdős, which is generous to a fault, comes along and offers 100 percent interest compounded continuously? Will you be "infinitely rich," as Erdős would say, at the end of one year? Well, not exactly. The amount you'll accumulate in one year is limited to $e$ dollars—in other words, $2.718. . . .

Gauss's great insight into the distribution of prime numbers was to see that as they got bigger and thinned out, their density was inversely proportional to the natural logarithm. According to the Prime Number Theorem—which Gauss conjectured in the late 1790s, and Erdős and Selberg proved by elementary methods in 1949—the average distance between two consecutive primes near a given number $n$ can be approximated by the natural logarithm of $n$, and the approximation gets much better as $n$ gets larger. For $n = 100$, the natural logarithm of $n$ is about 4.6, so the Prime Number Theorem predicts that for numbers in the vicinity of 100, on average 1 in every 4.6 numbers is a prime. How good is the prediction? Well, of the fifty numbers from 75 to 125, nine of them are in fact prime (79, 83, 89, 97, 101, 103, 107, 109, and 113), which means that on average, 1 in every 5.5 numbers is a prime—not a bad correlation with the theorem. For very large values of $n$, as $n$ approaches infinity, the difference between the density given by the natural logarithm and the actual density goes to zero. Why primes, the building blocks of all integers, should be intimately linked to $e$, the constant of growth and decay, can only be answered by the SF Himself.

If mathematical success is measured by revealing deep

connections among ideas that on the surface don't seem to be related, Euler gets the prize. He is responsible for perhaps the most concentrated and famous formula in all of mathematics, which in one bold stroke ties together $\pi$, $e$ and $i$ (the imaginary number, the square root of $-1$) as well as the most basic whole numbers 0 and 1. For Euler recognized that if you raised the number $e$ to the power $\pi$ times $i$ and added 1, you'd get 0. Behold the sheer elegance, hieroglyphic beauty, and austere conciseness of Euler's formula $e^{\pi i} + 1 = 0$, which has as much appeal for mystics as it has for mathematicians.

Lost in the beauty and compactness of the formula $e^{\pi i} + 1 = 0$ is a long history, because the acceptance and understanding of numbers like $\pi$, $e$, and $i$ did not come easily to mathematicians. Nor did the acceptance of much simpler numerical concepts like zero, negative numbers, and nonrepeating, nonterminating decimals like the square root of 2. Erdős was indeed precocious when he discovered negative numbers at the age of four. His discovery is all the more remarkable in light of the fact that it took Western culture until the seventeenth century fully to accept negative numbers. The mathematicians of Euclid's day were comfortable with the positive integers, 1, 2, 3, and so on, and they were comfortable enough with fractions provided they were expressed (cumbersomely in our modern eyes) as unit fractions. But the ancient Greeks had no concept of zero or negative numbers. Aristotle even went so far as to question whether 1 was a number, because numbers measured pluralities and 1 was only a unity. The Greeks had no trouble doing subtraction, and ancient herdsmen, who presumably could count their flocks, had no trouble subtracting three cows from six cows, but they didn't take seriously the concept of minus three cows.

As Martin Gardner put it, "A cow from a cow leaves

nothing, but adding a negative cow to a positive cow, causing both to vanish like a particle meeting its antiparticle, seems as ridiculous as the old joke about the individual whose personality was so negative that when he walked into a party, the guests would look around and ask, 'Who left?' " As late as 1660, Blaise Pascal, the father of probability theory, thought it nonsense to call anything less than zero a number. Greek and Renaissance mathematicians certainly knew how to solve equations with negative numbers; it's just that they thought of such quantities as "fictitious" entities. The rise of capitalism helped make these entities real; ledgers of credit and debt, and red ink on the balance sheet, paved the way for Western culture to embrace negative numbers finally in the seventeenth century.

Throughout the Dark Ages, Western mathematics was held back by the strange preference for unit fractions and an antiquated reliance on Roman numerals. Some Western mathematicians knew there was a better way, but their observations were shouts in the dark. Leonardo Fibonacci, despite his flirtation with unit fractions, was one who saw the light. Fibonnaci was born in the Italian city-state of Pisa late in the twelfth century, the son of a wealthy merchant and community leader. In Pisa, he learned Latin and studied the work of Euclid and other Greek mathematicians. When he was still a schoolboy, he moved to the Muslim city of Bugia, in North Africa, where his father had become a customs official, examining leather and furs before they were shipped back to Pisa. Young Leonardo got an education in Arabic culture, traveling around the Mediterranean, to Constantinople, Egypt, and Syria. He recognized that the Hindu-Arabic numerals, the numerals we use today, were superior to the Roman numerals he grew up with in the West.

Roman numerals worked just fine for addition and subtraction. For example, the sum of 10 and 3 is as easy to express with Roman numerals:

$$
\begin{array}{r}
\text{X} \\
+\ \text{III} \\
\hline
\text{XIII}
\end{array}
$$

as it is with Hindu-Arabic numerals:

$$
\begin{array}{r}
1\ 0 \\
+\ 3 \\
\hline
1\ 3
\end{array}
$$

To add Roman numerals, you simply group the symbols together. The most you have to do is to replace a bunch of lower symbols with a higher symbol, say substituting one V for five I's, as happens when you add 3 and 4:

$$
\begin{array}{r}
\text{III} \\
+\ \text{IIII} \\
\hline
\end{array}
$$

IIIIIII becomes VII

But multiplication is extremely cumbersome because the Roman numeral system lacks the idea of *place value*, or positional notation, which is second nature to our number system. When we write 23, the positions occupied by the 2 and the 3 are of crucial importance. We mean 2 tens because the 2 is in the tens' place and 3 ones because the 3 is in the ones' place. That's what makes it easy to multiply, say, 23 times 4:

$$
\begin{array}{r}
4 \\
\times\ 23 \\
\hline
92
\end{array}
$$

Think about how you actually do this. You multiply 3 ones by 4 ones to get 12 ones. The 12 ones is really 1 ten and 2 ones. So you write 2 in the ones' place and carry the 1 to the tens' place, holding it for a moment in your mind. Now you multiply the 2 in the tens' place by the 4 in the ones' place to get 8 in the tens' place. Adding the 8 to the carried 1 gives you 9 in the tens' place. *Voilà!*

Where do you even start if you're trying this with Roman numerals?

$$
\begin{array}{r}
IV \\
\times\ XXIII \\
\hline
\end{array}
$$

The answer is that no one probably tried. In thirteenth-century Pisa, multiplication was done on an abacus, in which different rows of sliding beads in effect slid in the concept of place value through the back door. An abacus worked fine except that there was no record of the steps in the computation, no way of saving your work.

Mathematicians in India in the sixth century had developed a place-value system and introduced the concept of a zero to keep their symbols in their proper places. Thus, a 1 with a 0 after it, or 10, is a very different number from a 1 alone. Erdős, who always joked that he was old and stupid, said the Indians were very clever, not just in their discovery of zero, but in their choice of similar-sounding Hindi words for stupid person (*buddhū*) and old person (*buddha*).

In the seventh century, Hindu scholars introduced Islam to the Indian number scheme, and the ideas of zero and place value spread rapidly throughout the Arabic world. Six centuries later, Fibonacci was so impressed with the ease of the Hindu-Arabic numerals that he wanted to make Pisan merchants aware of them. In 1202, he wrote *Liber abaci (Book of the Abacus)*, which, despite the title, had little to do with the abacus and a lot to do with liberating computations from the yoke of Roman numerals. The book seems quaint from the vantage point of the twentieth century, because it explains what we take for granted. "The nine Indian figures are: 9 8 7 6 5 4 3 2 1," the book begins. "With these nine figures, and with the sign zero . . . any number may be written."

To Fibonacci's chagrin, *Liber abaci* was ignored by the Pisan trading class, who, wallowing in prosperity, could not be bothered to adopt zero and renounce Roman numerals. The book got a better reception among Fibonacci's fellow mathematicians, and slowly, over time, it became the most influential work in getting the West to convert to Hindu-Arabic numerals. By the fifteenth century, the numerals were showing up on coins and gravestones; and by the seventeenth century, Western mathematics, which had fully emerged from the stagnation of the Dark Ages, was flourishing, thanks in no small part to zero, Hindu-Arabic numerals, and negative numbers.

In the seventeenth century, Western mathematicians also grappled head-on with the infinite—one of Erdős's favorite ideas—a concept that theretofore had been steeped in mysticism and largely avoided. God's power was thought to be infinite. Woe to the mere mortal who tried to reduce God's power to a symbol in an equation. What if He didn't like the representation?

Isaac Newton and Gottfried Wilhelm Leibniz dared to tread in this divine domain, bringing infinity into mainstream mathematics through their independent invention of *calculus*. Calculus is a powerful set of techniques for quantifying instantaneous change and rates of change, in anything from the motion of a planet to the flow of heat in an oven. Calculus involves both the infinitely small and the infinitely large. The infinitely small comes into play in "integration," the process of finding "the area under a curve" by dividing that area into smaller and smaller parts—*infinitesimals*, they were called—and summing the areas of each of these constituent parts. The infinitely large arises in the fundamental concept of calculus known as the limit. Consider the so-called geometric series $1/2 + 1/4 + 1/8 + 1/16 + 1/32 + \ldots + 1/2^n + \ldots$, where each denominator is double its predecessor. No matter how many finite terms you choose to sum in this series, you'll never get the number 1, although you can get as close to 1 as you want. But in the limit, where the number of terms approaches infinity, the sum is exactly 1.

Now consider the so-called harmonic series, consisting of the sum of the reciprocals of the positive integers:

$$1/1 + 1/2 + 1/3 + 1/4 + 1/5 + 1/6 + 1/7 + 1/8 + \ldots + 1/n + \ldots$$

What is the limit of the harmonic series as $n$ approaches infinity? The harmonic series doesn't look all that different from the geometric series. And yet while the geometric series converges to 1 in the limit, the harmonic series refuses to converge. Even though successive terms get increasingly smaller, the sum of the harmonic series can be made to exceed any positive number whatsoever, albeit only with major effort because the series grows so slowly. To exceed 5, for example, you need to sum the first 83 terms. To exceed 20, the first 300 million or so terms. To exceed 100, the first $10^{43}$ or so terms!

And then there are simple-looking infinite series that seem to misbehave entirely. Take a look at the series

$$1 - 1 + 1 - 1 + 1 - 1 + 1 - 1 + 1 - 1 + \ldots$$

What does it add up to? Grouping the terms one way,

$$(1 - 1) + (1 - 1) + (1 - 1) +$$
$$(1 - 1) + (1 - 1) + \ldots$$

the sum is definitely 0. But if the terms are grouped another way,

$$1 + (-1 + 1) + (-1 + 1) +$$
$$(-1 + 1) + (-1 + 1) + \ldots$$

the sum is definitely 1.

Since 0 obviously does not equal 1, something peculiar was going on. "Most people felt there was nothing illegal about regrouping the terms," said Graham. "After all, it was just addition. But the $0 = 1$ result showed that when it came to infinite series, it was a little tricky out there. People were confused for quite a while." Niels Henrik Abel, the nineteenth-century Norwegian genius, was driven wild by these series. A year before his death at the age of twenty-seven, Abel declared: "The divergent series are the invention of the devil, and it is a shame to base on them any demonstration whatsoever. By using them, one may draw any conclusion he pleases, and that is why these series have produced so many fallacies and so many paradoxes. . . ."

Even when mathematicians succeeded in taming infinite series, the concept of infinity itself still held plenty of surprises. At the end of the nineteenth century, for instance, infinity proved to be not one concept but several. Georg Ferdinand Ludwig Philipp Cantor, a German mathematician born in 1845 in St. Petersburg, offered a startling demonstration that infinity came in different sizes, just as integers did. In Galileo's *Dialogues on Two New Sciences*, written two and a half centuries before Cantor, the great Italian scientist called attention to the one-to-one correspondence between the counting numbers (1, 2, 3 ... ) and their squares, even though intuitively there seemed to be far fewer squares than counting numbers:

| 1 | 2 | 3 | 4 | 5 | 6 | 7 | ... |
|---|---|---|---|---|---|---|-----|
| ↕ | ↕ | ↕ | ↕ | ↕ | ↕ | ↕ | |
| 1 | 4 | 9 | 16 | 25 | 36 | 49 | ... |

Galileo did not know what to make of this baffling correspondence, but Cantor did. He claimed the one-to-one correspondence should be taken at face value, that in fact there were just as many squares as there were counting numbers. By the same token, Cantor said, there were just as many *even* counting numbers as there were counting numbers overall:

| 1 | 2 | 3 | 4 | 5 | 6 | 7 | ... |
|---|---|---|---|---|---|---|-----|
| ↕ | ↕ | ↕ | ↕ | ↕ | ↕ | ↕ | |
| 2 | 4 | 6 | 8 | 10 | 12 | 14 | ... |

And just as many primes:

| 1 | 2 | 3 | 4 | 5 | 6 | 7 | ... |
|---|---|---|---|---|---|---|-----|
| ↕ | ↕ | ↕ | ↕ | ↕ | ↕ | ↕ | |
| 2 | 3 | 5 | 7 | 11 | 13 | 17 | ... |

Cantor concluded that all these infinite sets—of squares, even integers, and primes—were the same size, a size he called *aleph-null*, after the first letter of the Hebrew alphabet. Sets of size aleph-null were said to be *denumerable*, or *countably infinite*, because their members could be put into one-to-one correspondence with the counting numbers. Cantor then asked himself whether the infinite set of rational numbers—all fractions of the form $a/b$—was also of size aleph-null. The number of rationals would seem to be much greater, because in the narrow gap between 0 and 1, for instance, there were infinitely many rationals: 1/171, 16/17, 19/65, 1/5, 5/6, 1/8, 231/232, and the list goes on and on. But in the realm of the infinite, things are often not what they seem. Cantor put forward an ingenious proof that the rationals were also countably infinite, aleph-null once again.

Cantor began the proof by arranging the rational numbers in an infinite array. The first row contains in increasing order of magnitude all fractions with denominator 1, the second row all fractions with denominator 2, the third row, denominator 3, and so on:

$$
\begin{array}{ccccccc}
1/1 & 2/1 & 3/1 & 4/1 & 5/1 & \ldots \\
1/2 & 2/2 & 3/2 & 4/2 & 5/2 & \ldots \\
1/3 & 2/3 & 3/3 & 4/3 & 5/3 & \ldots \\
1/4 & 2/4 & 3/4 & 4/4 & 5/4 & \ldots \\
1/5 & 2/5 & 3/5 & 4/5 & 5/5 & \ldots \\
\ldots & \ldots & \ldots & \ldots & \ldots \\
\end{array}
$$

Now, even though each row and column in this infinite array has infinitely many entries, Cantor showed that you still could count the fractions. In his ingenious "diagonal argument," he showed that the fractions could be put into one-to-one correspondence with the counting numbers by taking a diagonal tour through the array: one step to the right, then diagonally down to the left as far as you can go, then one step down, then diagonally up to the right as far as you can go, then repeat this procedure ad infinitum. (He knew, of course, that some fractions would be repeated; for example, 1/1, 2/2, 3/3, and 4/4 are all different representations of 1. When you come across alternative representations, just skip over them, he urged.)

$$
\begin{array}{ccccc}
1/1 \rightarrow 2/1 & 3/1 \rightarrow 4/1 & 5/1 \rightarrow \ldots \\
\swarrow \quad \nearrow & \swarrow \quad \nearrow & \swarrow \\
1/2 \quad 2/2 & 3/2 \quad 4/2 & 5/2 \ldots \\
\downarrow \quad \nearrow & \swarrow \quad \nearrow & \swarrow \\
1/3 \quad 2/3 & 3/3 \quad 4/3 & 5/3 \ldots \\
\swarrow \quad \nearrow & \swarrow \quad \nearrow \\
1/4 \quad 2/4 & 3/4 \quad 4/4 & 5/4 \ldots \\
\downarrow \quad \nearrow & \swarrow \quad \nearrow & \swarrow \\
1/5 \quad 2/5 & 3/5 \quad 4/5 & 5/5 \ldots
\end{array}
$$

$\cdots \quad \cdots \quad \cdots \quad \cdots \quad \cdots$

"On each diagonal you're looking at all the fractions where the numerator and denominator have a fixed sum," said Graham. It was clear to Cantor that by this diagonal tour he had achieved a one-to-one mapping between fractions and the counting numbers:

| 1 | 2 | 3 | 4 | 5 | 6 | 7 | 8 | 9 | ... |
|---|---|---|---|---|---|---|---|---|-----|
| ↕ | ↕ | ↕ | ↕ | ↕ | ↕ | ↕ | ↕ | ↕ | |
| 1/1 | 2/1 | 1/2 | 1/3 | 3/1 | 4/1 | 3/2 | 2/3 | 1/4 | ... |

Cantor himself was surprised by his proof that the rationals were countably infinite. "I see it," he said, "but I don't believe it!"

Cantor was also taken with the weird arithmetic of aleph-null, $\aleph_0$, the countably infinite. For example,

$$\aleph_0 + 1 = \aleph_0 \text{ and } \aleph_0 + \aleph_0 = \aleph_0$$

These strange properties—that when you add 1 to infinity you still have infinity and, for that matter, when you double infinity you still have infinity—were central to the paradox of Hotel Hilbert, named for the legendary German mathematician who heaped praise on Cantor's work. "From the paradise created for us by Cantor," said David Hilbert, "no one will drive us out." Hotel Hilbert was said to have an infinite number of rooms. On a particular night all the rooms were occupied, but a VACANCY sign hung outside. A potential guest arrived, and the desk clerk gave him the key to room 1, after asking the occupant of room 1 to move to room 2, the occupant of room 2 to move to room 3, the occupant of room 3 to move to room 4, and so on. The next night an infinite number of new guests arrived and none of the old guests had checked out. "No problem," the desk clerk announced. He moved the occupant of room 1 to room 2, the occupant of room 2 to room 4, the occupant of room 3 to room 6, the occupant of room 4 to room 8, the occupant of room 5 to room 10, in general the occupant of room $n$ to room $2n$. That freed up infinitely many odd-numbered rooms for the infinitely many new arrivals.

As if the arithmetic of infinity wasn't paradoxical enough, Cantor set out to find an infinite set that was bigger than the infinity of counting numbers. He investigated the set of real numbers, which, you'll recall, consists of all numbers that can be represented as decimals. Just start listing all the reals between 0 and 1 in no particular order, Cantor urged, and match them up with the counting numbers:

| | |
|---|---|
| .12146789 . . . | 1 |
| .32769234 . . . | 2 |
| .71234568 . . . | 3 |
| .43567233 . . . | 4 |
| .645946784 . . . | 5 |
| . . . | . . . |
| . . . | . . . |

It might seem that there should be enough counting numbers to match up with every real. But the reals simply can't be put into one-to-one correspondence with the counting numbers. Assume the list does contain all the reals. According to Cantor's diagonal argument, construct a new number from the digits in boldface, .12264 . . . , by replacing every single digit by a different digit of your choosing. For example, you could add one to each digit to produce the number .23375 . . . This new number is clearly a real number so it should appear somewhere on the list. But it can't be anywhere on the list because, by its very construction, its first digit differs from the first number on the list, its second digit differs from the second number on the list, its third digit differs from the third number on the list, and so on. Thus the list of reals isn't complete; there are more reals than counting numbers to pair them with. Cantor had found a larger infinity, which he called *aleph-one*.

"I entertain no doubts as to the truth of the trans-finites," he said, "which I recognized with God's help." Cantor's timing could not have been more fortuitous. No sooner had he published his proof than Pope Leo XIII delivered an influential encyclical calling on Catholics to open their minds to the teachings of science. With the pope as his unwitting publicist, Cantor had an audience. "From me," he boasted, "Christian Philosophy will be offered for the first time the true theory of the infinite." Though some Christian thinkers were critical of his trespassing on God's territory, others took him at his word when he said he had shown God's power to be even greater than ordinarily assumed. God's realm, he argued, was not merely the countably infinite but the even greater transfinite. In a bow to Cantor, Erdős described the Book as having a transfinite number of pages.

Even with a direct line to God, Cantor did not have an easy life. His closest son mysteriously dropped dead four days before his thirteenth birthday, and Cantor himself was in and out of mental institutions, fighting breakdowns and depression so debilitating that he'd sit rigid and mute for days at a stretch. In this trancelike state he'd hear the voice of God sharing pages from the Book. When he was up and about, he had a soft spot for conspiracy theories. He got caught up in the intellectual circus of trying to prove that Francis Bacon wrote the plays of Shakespeare and claimed to decipher messages about the first king of England, "which will not fail to terrify the English government as soon as the matter is published." Cantor spent the last year of his life surviving on wartime rations in a mental hospital in Halle, Germany, pleading with his family to come and take him home. He died there of heart failure, on January 6, 1918, at the age of seventy-three.

Cantor's reception among his fellow mathematicians was as mixed as it was among theologians. Hilbert, to be sure, poured on the praise, as did Bertrand Russell, who described Cantor's insights into the infinite as "probably

the greatest of which the age can boast." But Leopold Kronecker, a pillar of the German mathematical community,
dismissed the transfinite, issuing his oft-quoted rebuke:
"God made the integers; all else is the work of Man." Kronecker meant that only the integers were real, that other
kinds of numbers were just figments of mathematicians'
hyperactive imaginations. Henri Poincaré, the great French
geometer, echoed Kronecker and suggested that future
generations of mathematicians would look back on Cantor's work as "a disease from which one has recovered."
The severity of the criticism exacerbated Cantor's mental
condition.

In his more lucid moments, Cantor had wondered
whether there was an infinity larger than the counting
numbers but smaller than the reals. He believed that there
wasn't such an infinite set but he was unable to prove this
conjecture, called the Continuum Hypothesis. In perhaps
the most famous speech in mathematics, at the Second International Congress of Mathematicians held in Paris in
1900, Hilbert posed twenty-three problems that he said
cried out for solution in the new century. First on his list
was the proof of the Continuum Hypothesis. Long after
Cantor's death, Gödel also tried to prove the Continuum
Hypothesis but he, too, was defeated. In 1963, Gödel's former assistant, a twenty-nine-year-old named Paul Cohen,
stunned the mathematics community with a proof that the
Continuum Hypothesis could never be proved by the mathematical axioms in common use. Cohen showed that you
could assume the Continuum Hypothesis to be either true
or false—take your choice—without contradicting other
results about infinite sets. Cohen's proof, though, was not
easy for his colleagues to understand.

"Mathematicians knew there was only one way to
know whether the proof was correct," John Barrow wrote

in *Pi in the Sky*, "and so soon Cohen found himself knocking on the door of Gödel's residence in Princeton." Gödel, who was in the throes of paranoia those days, didn't let his old assistant in. He "opened the door just wide enough for Cohen's proof to pass, but not so wide as to allow Cohen to follow. But two days later Cohen received an invitation to tea with the Gödels. The proof was correct. The master had given his *imprimatur*." Gödel's incompleteness theorem of 1929 had finally reared its ugly head in a real situation. When Gödel put forward his theorem, many mathematicians reluctantly acknowledged its truth but felt that only very contrived mathematical statements would prove to be undecidable. But now Cohen had shown that the problem Hilbert most wanted to solve was undecidable.

Erdős never fully made his peace with Cohen's result and the counterintuitive work of his friend Gödel. "If I were alive in a thousand years, I would ask," said Erdős, "has there been a solution of the Continuum Hypothesis? Suppose you have an infinite intelligence, could it decide whether the Continuum Hypothesis is true or false? Most of the logicians believe it cannot be done. Yes, the Continuum Hypothesis is in a way undecidable. But it is conceivable that an infinite intelligence could decide it, because it is conceivable that there would be methods of proof which we can't understand but which a higher intelligence would understand. I don't want to say that such methods exist. I wouldn't even say that I believe that such methods exist, but they might exist." There is an old joke Erdős liked to tell about a man who is trying to convert people who asks, "What would you say to Jesus if you saw him on the street?" Erdős said he'd ask Jesus if the Continuum Hypothesis was true. "And there would be three possible answers for Jesus," Erdős said. "He could say, 'Gödel and Cohen already taught you everything which is to be known about it.' The second answer would be, 'Yes, there is an answer but unfortunately your brain isn't sufficiently developed yet to know the answer.' And Jesus could give a third answer: 'The Father, the Holy Ghost, and I have been thinking about that long before creation, but we haven't yet come to a conclusion.' Perhaps this is the nicest answer. But most logicians would accept the first answer that Gödel and Cohen told you everything there is to know about the Continuum Hypothesis."

An infinite intelligence, said Graham, would not be of any help here. Cohen showed that there is no ultimate sense in which the Continuum Hypothesis is either true or false. Either possibility is consistent with the rest of mathematics. "We have a certain set of axioms that we use in mathematics," said Graham, "and a certain set of rules for how you manipulate the axioms to produce new theorems. And if you add the Continuum Hypothesis—that is, if you add the assertion that there is no infinite set larger than the counting numbers and smaller than the reals—you won't run into any more trouble than you would have otherwise.

"Now, the scary thing is that nobody knows in fact whether the current axioms we're using are actually consistent. It may turn out that someone will find a way to prove some result and also prove that the result doesn't hold. That makes you worry a little bit. When we do mathematics, are we just playing some crazy game that really doesn't make any sense at all because the axioms we're assuming are contradictory? Once you have a system in which you can prove $X$ is true and also prove $X$ is false, then you can prove anything. So what are you doing?

"There's an old Robert Mankoff cartoon in which a thoughtful guy who looks like a mathematician only dressed better is in a restaurant discussing his bill with the waiter. 'The arithmetic seems correct,' he says, 'yet I find myself haunted by the idea that the basic axioms on which the arithmetic is based might give rise to contradictions that would then invalidate these computations.' I suppose we can take consolation in the fact that if mathematics were contradictory, the contradictions should have shown up by now. They haven't, so mathematics is probably okay. But the trouble is that set theory went happily along for some time before the ominous paradoxes arose that defeated Frege and Russell."

For all his waffling about Gödel and Cohen, Erdős was quick to embrace Cantor's discovery of the transfinite. Erdős was always trying to extend finite combinatorial problems into the realm of the infinite and beyond, and he made fundamental contributions to the theory of "inaccessible cardinals," infinite sets much larger than the real numbers.

One beautiful result of Cantor's work was a proof that transcendental numbers (irrational numbers that can never be the solutions to ordinary algebraic equations) actually exist. As ubiquitous as $\pi$ and $e$ had become in mathematics, before the nineteenth century no one had proved that these numbers were transcendental. For all mathematicians knew, they might someday discover an ordinary equation that $\pi$ or $e$ solved. It wasn't until 1873, the same year Cantor showed the existence of the transfinite, that Charles Hermite was able to prove the transcendence of $e$, and the proof thoroughly exhausted him. "I shall risk nothing," Hermite told a colleague, "on an attempt to prove the transcendence of $\pi$. If others undertake this enterprise, no one will be happier than I at their success, but believe me, my dear friend, this cannot fail to cost them some efforts."

In 1874, Cantor showed that the set of transcendental numbers was too big to be countable. What Cantor had done, without actually demonstrating that any particular number was transcendental, was to show that the seemingly rare transcendental numbers were not rare at all. In fact, they were infinitely more numerous than the familiar integers. His proof gave no clue as to how to construct a single one of the multitudes of transcendentals. And they were hard to find, despite Cantor's demonstration that they outnumbered the counting numbers. Eight more years would go by before Ferdinand Lindemann of the University of Munich finally proved the transcendence of $\pi$. That was in 1882, more than 2,100 years after Archimedes made the first crude estimates of the value of $\pi$, approximating it to 2 decimal places. Today $\pi$ is known to more than 50 billion decimal places. In the engineering world, you need know $\pi$ to only 39 places in order to compute "the circumference of a circle girdling the known universe with an error no greater than the radius of a hydrogen atom."

The idea of an existence proof—demonstrating the existence of something without being able to display that something—was unsettling to mathematicians of the old school. In time this method of proof was pretty much fully accepted. As well it should have been. Existence proofs, to be sure, aren't at all troubling in everyday, nonmathematical life: If I manage a stadium, for instance, that has 50,000 seats, and I count 49,999 people filing through the gates for an open-seating concert, I can be sure of the existence of an empty seat, even if I don't know precisely where it is.

One of Erdős's most fundamental contributions to mathematics was to come up with a powerful new form of an existence proof called the *probabilistic method*. Erdős introduced the technique in 1947 to solve a Ramsey problem. Imagine a group of people at a party. Erdős in effect flipped a coin to decide whether each pair of people are friends or strangers. The result is a random pairing of friends and strangers. Erdős then proved that the probability of avoiding a particular mix of people is overwhelmingly good provided the party isn't too large. "If what you're dealing with gets too large," said Graham, "you can't avoid structure—that's Ramsey theory. With his probabilistic method, Erdős proved that you can still get quite large and avoid a particular structure, although his method doesn't tell you how to actually avoid the structure. It's like certain methods that prove a number is composite but give you no hint as to what the prime factors are. You know the factors must exist, but you have no clue. It's the same with the probabilistic method. And there are a number of things that are known to exist by this probabilistic method but people have no idea how to build them."

The idea of flipping a coin to help solve mathematical problems was disconcertingly radical in a field prized for its precision, but it is now commonplace in computer science. The irony is that Erdős, who shunned computers, made a major contribution to the theory of computing. Making a random choice, it turns out, often proves to be an excellent way of avoiding computational gridlock. "Two people walking in opposite directions on a sidewalk both step to the same side to avoid colliding, then do it again and again," Ivars Peterson wrote in *The Jungle of Randomness*. "They get trapped in an embarrassing dance until the stalemate is somehow finally broken. A coin toss can solve the problem for both the inadvertent dancers and the computer that is engaged in resolving conflicting instructions or deciding which of two or more equally likely courses to take."

"Random methods were effective in the routing of telephone calls," said Graham, who spent thirty-five years in R&D at AT&T. "If the system got overloaded on a regular link, it routed the call to another link. Suppose that normally a call between New York and Chicago took link 1. And there was a rule that said if link 1 was busy, go to link 2, and if that was busy, go to link 3, and so forth. Well, if you were too specific like this, the traffic got all piled up. You want to avoid systematic bias. It was better to have no rule and just pick a random link. Things work a little differently today, but the random approach had its heyday at the phone company.

"The same computational challenge came up in the early days of Star Wars. You have all these incoming missiles and you're defending things. Well, how do you coordinate the counterattack? Who shoots down what? The idea is to do it at random. The first pass is to look out there and pick a random missile. If you have lots of defensive missiles randomly attacking even thousands of missiles coming in, it's very unlikely that you're not going to pick them all off. Of course, the trouble is there's a penalty if you're wrong. Oops, no one got that one. Washington up in smoke! You can prove that the random defense is very effective, but that wasn't convincing. Issues of national defense defy logic. It doesn't matter what you can prove or not prove."

In 1989, in celebration of Erdős's probabilistic method, mathematicians who had come together for the Random Graph Conference in Poznań, Poland, staged a random race. As they ran around a track, they didn't know how to pace themselves because they didn't know when the race was going to end. Before the race started, Erdős, as the master of ceremonies, tossed a huge die. This determined the initial number of laps. But as the mathematicians approached the finish line, Erdős threw the die again, specifying an additional number of laps. Erdős grinned broadly, clearly enjoying his role as the tormenting SF.

# 6

## GETTING THE GOAT

> My only advice is, if you can get me to offer
> you $5,000 not to open the door, take the money
> and go home.
>
> —Monty Hall

Although numbers were Erdős's intimate friends, he did
occasionally misjudge them. Good as he was, his intuition
was not always perfect. Indeed, the last time he visited
Vázsonyi, at his retirement home in California's wine country, he tripped up on a tricky brain teaser posed in "Ask
Marilyn," Marilyn vos Savant's column in *Parade* magazine. Flashy and confident, vos Savant is someone professional mathematicians love to hate. She bills herself as the
person with the "Highest IQ" ever recorded, a whopping
228, according to *The Guinness Book of World Records.* She
sports a wedding ring of pyrolytic carbon, a special material
used in the Jarvik artificial heart, which was invented by
her husband, Robert Jarvik. Her reputation in the mathematics community was not helped by her book *The World's
Most Famous Math Problem* (1993), in which she questions
Wiles's proof of Fermat's Last Theorem and Einstein's theory of relativity. "Ask Marilyn" has been described as a

kind of "Hints from Heloise" for the mind, with lots of mathematics thrown in. Some of the dislike for her stems from puzzle envy: Her *Parade* column is read by millions every Sunday, and the accompanying books and speaking engagements have earned her a good living. Many professional mathematicians, on the other hand, have not earned a cent from their books.

In her column for September 9, 1990, vos Savant answered a well-known brain teaser submitted by one of her readers. You're on a game show and you're given the choice of three doors. Behind one door is a car, behind the other two are goats. You choose, say, door 1, and the host, who knows where the car is, opens another door, behind which is a goat. He now gives you the choice of sticking with door 1 or switching to the other door? What should you do?

This was the so-called Monty Hall dilemma faced by guests on Monty Hall's classic TV game show *Let's Make a Deal*, only the consolation prizes weren't goats. Vos Savant advised her correspondent to switch doors. Sticking with the first choice gives a one-third chance of winning, she said, but switching doubles the odds to two-thirds. To convince her readers, she asked them to imagine a million doors. "You pick door No. 1," she said. "Then the host, who knows what's behind the doors and will always avoid the one with the prize, opens them all except door No. 777,777. You'd switch to the door pretty fast, wouldn't you?"

Evidently not. No sooner had her column appeared than she was besieged by mail from readers who disagreed, including many mathematicians. They maintained the odds were only fifty-fifty, not two-thirds, in favor of switching. In her December 2, 1990, column vos Savant ran some of the letters:

As a professional mathematician, I'm very concerned with the general public's lack of mathematical skills. Please help by confessing your error....

Robert Sachs, Ph.D., George Mason University

You blew it, and you blew it big! I'll explain: After the host reveals a goat, you now have a one-in-two chance of being correct. Whether you change your answer or not, the odds are the same. There is enough mathematical illiteracy in this country, and we don't need the world's highest IQ propagating more. Shame!

Scott Smith, Ph.D., University of Florida

This time, to drive her analysis home, vos Savant made a table that exhaustively listed the six possible outcomes:

| Door 1 | Door 2 | Door 3 | Outcome *(choose No. 1 and stick with No. 1)* |
|--------|--------|--------|---------|
| Car | Goat | Goat | Win |
| Goat | Car | Goat | Lose |
| Goat | Goat | Car | Lose |

| Door 1 | Door 2 | Door 3 | Outcome *(choose No. 1 and switch)* |
|--------|--------|--------|---------|
| Car | Goat | Goat | Lose |
| Goat | Car | Goat | Win |
| Goat | Goat | Car | Win |

The table demonstrates, she wrote, that "when you switch, you win two out of three times and lose one time in three; but when you don't switch, you only win one in three times."

But the table did not silence her critics. In a third column on the subject (February 17, 1991), she said the thousands of letters she received were running nine to one against her and included rebukes from a statistician at the National Institutes of Health and the deputy director of the Center for Defense Information. The letters had gotten shrill, with suggestions that she was the goat and that women look at mathematical problems differently from men. "You are utterly incorrect about the game-show question," wrote E. Ray Bobo, a Ph.D. at Georgetown, "and I hope this controversy will call some public attention to the serious national crisis in mathematical education. If you can admit your error, you will have contributed constructively toward the solution to a deplorable situation. How many irate mathematicians are needed to get you to change your mind?"

"When reality clashes so violently with intuition," vos Savant responded in her column, "people are shaken." This time she tried another tack. Imagine, she said, that just after the host opened the door, revealing a goat, a UFO lands on the game-show stage, and a little green woman emerges. Without knowing what door you originally chose, she is asked to choose one of the two unopened doors. The odds that she'll randomly choose the car are fifty-fifty. "But that's because she lacks the advantage the original contestant had—the help of the host. . . . If the prize is behind No. 2, the host shows you No. 3; and if the prize is behind No. 3, the host shows you No. 2. So when you switch, you win if the prize is behind No. 2 *or* No. 3. *YOU WIN EITHER WAY!* But if you *don't* switch, you win only if the prize is behind door No. 1." Vos Savant was completely correct, as mathematicians with egg on their faces ultimately had to admit.

■

Vázsonyi told Erdős about the Monty Hall dilemma. "I told Erdős that the answer was to switch," said Vázsonyi, "and fully expected to move to the next subject. But Erdős, to my surprise, said, 'No, that is impossible. It should make no difference.' At this point I was sorry I brought up the problem, because it was my experience that people get excited and emotional about the answer, and I end up with an unpleasant situation. But there was no way to bow out, so I showed him the decision tree solution I used in my undergraduate Quantitative Techniques of Management course." Vázsonyi wrote out a "decision tree," not unlike the table of possible outcomes that vos Savant had written out, but this did not convince him. "It was hopeless," Vázsonyi said. "I told this to Erdős and walked away. An hour later he came back to me really irritated. 'You are not telling me *why* to switch,' he said. 'What is the matter with you?' I said I was sorry, but that I didn't really know why and that only the decision tree analysis convinced me. He got even more upset." Vázsonyi had seen this reaction before, in his students, but he hardly expected it from the most prolific mathematician of the twentieth century.

"Physical scientists tend to believe in the idea that probability is attached to things," said Vázsonyi. "Take a coin. You know the probability of a head is one-half. Physical scientists seem to have the idea that the probability of one-half is fused with the coin. It's a property. It's a physical thing. But say I take that coin and toss it a hundred times and each time it comes up tails. You will say something is wrong. The coin is false. But the coin hasn't changed. It's the same coin that it was when I started to toss it. So why did I change my mind? Because my mind has been upgraded with information. This is the Bayesian view of probability. It took me much effort to understand that probability is a state of mind. My hypothesis is that Erdős had this idea of probability as being attached to physical things and that's why he couldn't understand why it made sense to switch doors."

Vázsonyi's retirement home is on the edge of a golf course. As Erdős pondered the Monty Hall problem and thought about the SF only knows how many other conjectures, "he insisted on walking along the golf course," said Laura Vázsonyi, "which they tell you not to do. You can be hit by balls; they're like bullets. There are signs everywhere warning against walking along the golf course. And we tried to stop him. But he wasn't too easily dissuaded. He never wanted his freedom to be restrained in any way."

When Erdős returned from his walk, Vázsonyi tackled the Monty Hall dilemma with a technique developed by Erdős's late friend Stan Ulam, called the *Monte Carlo method*. In 1946, Ulam had played a lot of solitaire while recovering from his bout with encephalitis. "After spending a lot of time trying to estimate [the odds of particular card combinations] by pure combinatorial calculations, I wondered whether a more practical method than 'abstract thinking' might not be to lay [the cards] out say one hundred times and simply observe and count the number of successful plays," Ulam recalled. "This was already possible to envisage with the beginning of the new era of fast computers, and I immediately thought of problems of neutron diffusion and other questions of mathematical physics, and more generally how to change processes described by certain differential equations into an equivalent form interpretable as a succession of random operations." The idea of this Monte Carlo method, named in honor of a relative of

Ulam's who was always sneaking off to the roulette wheels at Monte Carlo, was to compute the odds of an event not by solving equations but by randomly simulating the event on a computer. In other words, to find the odds of a particular card formation in solitaire, ask a computer to deal hundreds of random hands, and count the incidence of the formation in question.

On his PC Vázsonyi ran a Monte Carlo simulation of the Monty Hall dilemma. Erdős, who never had much use for computers, watched the PC randomly choose whether to switch or stick. The outcome of hundreds of trials favored switching two to one, and Erdős conceded that he was wrong. But the simulation was no more satisfying than the computer proof of the Four Color Map Theorem. It wasn't the Book proof. It didn't reveal why it was better to switch. Erdős, who found Vázsonyi's explanations lacking, was ready to leave.

This was the last time Vázsonyi and Erdős would see each other. Erdős wanted Vázsonyi to drive him two hours to San Francisco International Airport.

"I'm not going to do that," Vázsonyi told him. "It's too far."

"How will I get there?" Erdős said.

"I'll put you on a bus."

"That won't work."

"Why?"

"When I get off the bus, what am I going to do?"

"When you get off the bus, the driver will take your suitcase and put it on the sidewalk. There will be a nice man in front of the counter—United Airlines. You show the guy your ticket and he'll take your bag."

Erdős wasn't happy about going to the airport by himself but finally consented. "He called later," said Vázsonyi, "and said it worked out just fine. He was so used to being taken care of. He assumed that there was nothing he could do alone. From his early upbringing by his mother, he was put in this frame of mind."

Erdős did not forget the Monty Hall problem. He called Graham and demanded the Book proof. "The key to the Monty Hall problem," Graham said, "is knowing ahead of time that the host is always going to give you the chance to pick another door. That's part of the rules of the game which you have to figure into your thinking." Erdős accepted Graham's explanation.

"When he didn't understand something," said Graham, "he did not make it easy for you to convince him. He was always interrupting and getting angry. Conversely, when he tried to explain a proof to you, it was not always easy to follow him. Once Fan was so annoyed with him that she said she was never going to work with him again. I was always the intermediary between the two of them, part translator, part peacemaker. Paul left out a lot of mental steps when he was explaining something. If you were with him a lot, you could stop him and make him fill them in. Fan does the same thing when she explains something, only she leaves out different steps. So when they're communicating back and forth, they're leaving out too many steps. Paul had a problem he really cared about and Fan solved it. He asked her to explain the solution. She had barely started when he suggested another approach. She said, 'No, Paul, I'm explaining this.' Finally after half an hour she was completely frustrated. And he said, 'I don't think Fan's English is good enough to explain this to me.' That sent her over the edge because her English is perfectly good. Paul just wanted to see it in his way. When you've thought about a complicated argument and you have your own notation in your mind, and Paul interrupts and says, 'Well, try it this other way and define such-and-such as the inverse,' it's like asking a right-handed person to do a difficult new athletic stunt left-handed. 'What? I can barely do it right-handed.'

"Near the end of his life Paul appreciated that his explanations were sometimes hard to follow. He realized this when he looked back at his old papers and was impressed by how hard it was for him to understand his own arguments of thirty or forty years earlier. He asked me once, 'Do you notice anything different about me?' I said, 'No, not really.' He said, 'You're supposed to be so perceptive, haven't you noticed how depressed I am lately?' I said, 'Well, maybe.' 'I see now,' he said, 'that I'm really losing it. I don't know what to do about it. It's pretty depressing.' I said, 'Paul, you're still way up above everyone else.' He used to ask me to check things because he was afraid of making a stupid mistake, but he usually didn't make mistakes." Which was just as well, owing to his often-repeated comment that if he could no longer do mathematics, he would have to commit suicide.

Time and again, Erdős's wry sense of humor lifted his own spirits. When he overlooked a simple result in graph theory once, he said, "I should be put under tutelage.... Hungary used to be a semi-feudal country and if a rich aristocrat became old and spent all his money on poison, bosses, and noise, his family put him under tutelage. He was given a large sum of money but couldn't touch his main capital."

Erdős's tired and sickly appearance deceived his friends over the years. Back in the 1940s, Vázsonyi recalled, "we always thought that his health was so fragile that he was not going to live long. He looked so frail. He always looked unhealthy. But he outlived most everybody." It was only during the last ten years of his life that health problems slowed Erdős down a bit, although he continued to do work at what by any other mathematician's standard was a frenetic pace. He began to lose sight in one eye, but didn't want to take time out from mathematics to get the requisite care.

"He dropped his glasses once," said Ralph Faudree, a Ramsey theorist at the University of Memphis, "and one of the lenses shattered but didn't fall out. Erdős said, 'No problem. I can't see out of that eye anyway,' and started to put the glasses back on. I was afraid that shards of glass would get in his eye, so I popped the broken lens completely out. I knew Graham had a duplicate pair at Bell Labs, so we called him and got them sent down. A little later I saw Paul wearing a pair of glasses. I didn't think he could have gotten them that fast. 'Are those from Graham?' I asked. 'No,' he said, looking puzzled. 'These don't help much either.' He thought for a moment and then added, 'I must have picked up someone else's glasses.' He thought some more, and then burst out smiling: 'It's probably the guy I'm staying with! I'll call him and see.' "

Eventually he lost all vision in that eye and badly needed a corneal transplant. A suitable donor was hard to come by. Ralph Faudree's wife, Pat, who is a nurse, pulled some strings to get him advanced in the queue. ("Surgery will advance all of mathematics," she told the hospital.) Still months passed before the Faudrees got a call that a match had been found. Erdős had just left Memphis and was on a plane to Cincinnati, where he was scheduled to give a talk. At first he refused to skip the talk and get the transplant. It took lots of stern prodding to get him back.

The transplant took about two hours. Before the operation, the doctor carefully explained to him the procedure.

"Doctor," Erdős said, "will I be able to read?"

"Yes," said the doctor. "That's the whole point of the surgery."

Erdős went into the operating room, and when the lights were dimmed, he immediately got agitated. "Why are you turning the lights down?"

"So we can do the surgery."

"But you said I'd be able to read."

He then had a huge argument with the surgeon about why, since only one eye was being deadened, he couldn't read a mathematics journal with the other, good eye. The surgeon made a series of frantic calls to the Memphis math department. "Can you send a mathematician over here at once so that Erdős can talk math during surgery?" The department obliged, and the operation went smoothly.

"After a corneal implant, it takes some time for your brain to learn to focus both eyes together," said Faudree. "Erdős spent a miserable year, which he let everyone know about, seeing double. That was tough for him because reading mathematics was so important to what he did."

Erdős's heart wasn't doing much better than his eyes. "For the past decade he had trouble with a sudden, sharply elevated pulse," said Faudree. "I was eating with him in a Greek restaurant when he realized that his heartbeat had shot up to 150. I rushed him to the emergency room. I remember vividly seeing him hooked up to a monitor. You could read his elevated pulse. But suddenly it started dropping, not just dropping, plunging. I thought it might drop to zero but it bottomed out at 65. A cardiologist kept him in the hospital but Erdős insisted on working. He had papers spread out all over the hospital bed and a parade of mathematicians coming in and out. The nurses tried to restrain him but couldn't. He did the same thing when he was hospitalized in Hungary. He drove doctors and nurses crazy."

"My wife and I visited him in Budapest in the spring

of 1988," said Faudree's Memphis colleague, Chip Ord-man. "He had had a small heart attack and was in the hospital. He told us that they had to move him from one hospital to another because the first place didn't have a room big enough to hold all the visitors. Sure enough, his new room was huge, and it was a complete mess. There were journals piled up and papers everywhere. Erdős was lying there, holding three mathematical con-versations at once, in Hungarian with a group in one cor-ner, in German with a group in another corner, and in English with a third group. All this while talking to me and my wife. The doctors came in and he said, 'Go away! Can't you see I'm busy? Come back in a few hours!' And that's what they did."

Eye ailments and heart problems did not stop him from resuming his twenty-five-country lecture circuit. He observed that the audiences for his talks were growing to the point where to accommodate everybody he would need a larger lecture hall, but then his old and feeble voice would not carry. He speculated on the reason for his increased popularity: "Everyone wants to be able to say, 'I remember Erdős. Why I even attended his last lecture.' "

The last year of his life, 1996, was particularly rough. The annual International Symposium on Combinatorics, Graph Theory, and Computing was held in March around Erdős's birthday. The conference alternated between Boca Raton, Florida, and Baton Rouge, Louisiana, and in both cities he traditionally lectured on Thursdays. (In 1994, in Boca, he had complained about becoming "probably a square for the very last time," a reference to his reaching the age of nine squared, or eighty-one, and not expecting to make it to one hundred, or ten squared.) About halfway through his March 1996 lecture, also in Boca, he got up to write at the blackboard and suddenly fell down stiff as a board. His blood pressure was very low, his heart rate down to 37. He was lying there flat but with his microphone still attached. People were scared and anxious and security was trying to get them to file out. "Tell them not to leave," he said, regaining consciousness. "I have two more problems to tell them." Later that day, recalled his friend Alexander Soifer, "there was an excursion to New Orleans for Mardi Gras. Of course I offered to stay with him. But he said, 'Just go and enjoy yourself. I'm fine. It's nothing. After Mardi Gras, come to my room and we'll do some mathematics.' "

In June he was listening to Gerhart Ringel, a mathematician from Santa Cruz, deliver a talk at a conference in Kalamazoo, Michigan. "As the talk ended and people were leaving, Paul, who was sitting in the front row, quietly asked him a question," said Michael Jacobson, a set theorist at the University of Louisville who wrote two papers with Erdős. "In the middle of his question he fell over and was out cold. By late that morning surgeons had put in a pacemaker. Sure enough that very evening Paul attended the closing banquet. His two heart surgeons were sitting next to him. Paul got up, took a little bow, introduced the surgeons, and then said, 'Now I just want to finish asking Dr. Ringel my question.' That said it all: Mathematics was his life."

"He never wanted pity," said Soifer. "When he called me from Kalamazoo, he didn't mention that he had just received a pacemaker. Instead, he asked about my kids and my wife and questioned me about my European plans. I said I'd come to Budapest. And he said, 'No, I'll come to Prague.' He didn't say a word about his heart condition. I only heard about it later. It was very painful to learn in September—I was in Prague and we were about to meet up—that Erdős had passed away. He made us used to the idea that he would live forever. For decades he had been joking about old age and death. All his humor about old age was like an injection for us. We were vaccinated and prepared that Paul would live forever—or for at least as long as we did."

# 7

## SURVIVORS' PARTY

Dear Ron,
    Please send me Uncle Paul's birthdate if you have it. Then we will have something to look forward to and not be so depressed.

                                    Thanks, Ed

    P.S. I hope he is giving the SF all sorts of trouble.


Ed,
    I think Erdős is too busy reading the Book to give anyone much trouble (for a while). Paul's birthday is March 26, 1913.

                                    Ron


March 1997 at the University of Memphis: the students had scattered for spring break. The campus would be deserted if it weren't for 260 world-class mathematicians from four continents who had come together for the 919th meeting of the American Mathematical Society. Apple trees were blossoming, warblers were singing to their mates, and the muddy waters of the nearby Mississippi were surging to flood levels. Inside a boxy brick building, in windowless lecture rooms, immune to the explosion of spring, were the mathematicians, cheerfully holed up for two long days of proving and conjecturing.

Nothing about the design of this inelegant building gave a clue to the elegant thinking that went on inside. There were no pi signs etched in the bricks, no statues of Gauss or Euler in the courtyard. There was no relief of Archimedes drawing a circle in the sand or running half-naked from his bath shouting "Eureka!" And there was no wrenching painting of Hypatia of Alexandria, who met her death early in the fifth century at the hands of a mob who dragged her from her classroom and peeled off her skin with oyster shells.

What cried out "math department" was not the look of the place but the notices plastered haphazardly on the hallway walls. A Day-Glo orange poster advertised a seminar in ergodic theory; a more subdued notice billed a special lecture on quasiconformal mappings; another called for submissions to an upcoming graph theory conference. On the wall too were Erdős obituaries, random aphorisms ("Safety comes in numbers," "Mathematicians do it better"), and curious musings, including a passage from an unidentified source:

> In some countries it is the style of the university professor to stand at the lectern in the front of the room and to read the textbook to the class. Questions are considered a rude Americanism. An extreme example of a teaching style that is virtually orthogonal to what we Americans are used to is one that was used by the celebrated Hungarian analyst F. Riesz. He would come to class accompanied by an Assistant Professor and an Associate Professor who would act as his assistants. The Associate Professor would write the words on the blackboard. And Riesz would stand front and center with his hands clasped behind his back and nod sagely.

The cartoons on the walls also gave the place away:

"YOU WANT PROOF? I'LL GIVE YOU PROOF!"

But among all these mathematical postings was the odd Pentecostal flyer: "The Flesh of God. The Promise is Unto You. (The Holy Ghost evidenced by speaking in other tongues)." And on a blackboard covered with mathematical notation were the words, in large, childlike handwriting: "Jesus said to him: 'I am the way, the Truth and the Life. No one comes to the Father but through me.'" Here in Memphis, in the heart of the Bible Belt, the mathematicians weren't the only ones who were trying to discern what's on the SF's mind.

A man of slight build in his fifties was pacing in the hallway, rehearsing the talk he was scheduled to give later. "I'm walking testimony to the fact that brain cells atrophy if you don't use them," he said. "I've had so little time for mathematics these past few months, ever since I took on administrative responsibilities at my college, that I've become rusty. It would be embarrassing if I can't recount the difficult steps in my own proof. I should have brought my younger collaborators along to present our results. Erdős liked to joke that we lose 50,000 brain cells a day, and here I am proving him right."

On the wall behind the nervous pacer was a Sidney Harris cartoon that wouldn't give him comfort: An older mathematician is shown staring vacantly at a blackboard on which is written:

$$\begin{array}{r} 2 \\ + \ 4 \\ \hline \end{array}$$

Two other, younger mathematicians are observing him from an adjacent blackboard marked up with far more complex problems. One whispers to the other, "He was very big in Vienna."

Like chess grandmasters and ballet stars, strong mathematicians fear the day when they sense their abilities slipping away—and that day usually comes long before old age. As Erdős put it, quoting his friend Ulam, "The first sign of senility is that a man forgets his theorems, the second sign is that he forgets to zip up, the third sign is that he forgets to zip down."

The pacing man stopped abruptly and peered into one of the lecture rooms. A young graph theorist, who couldn't possibly be worrying yet about senility, was wrapping up a crowd-pleasing talk. He paused in his remarks and stared intensely at a long chain of symbols on the blackboard. After a minute or two, he shrugged his shoulders, turned back to his audience, and announced, "It's a little bit fishy to see that this is true, but if you think hard for thirty minutes, you'll know that it is true." Everyone laughed. His talk was over, but three of his listeners stayed in their seats, taking him up on the deep think.

The rest filed out and raced toward a folding table that supported two large brown plastic tanks, one labeled WITH CAFFEINE, the other WITHOUT CAFFEINE. The mathematicians had already tanked up once, before the graph theorist's talk, and now the labels on the tanks were curiously switched. The mathematicians were in a tizzy and stared, puzzled, at the tanks. Their collective brainpower, fine for discerning the latest wrinkles in infinite graph theory, was stalling out on the case of the true Joe.

Finally a lightbulb went off in the head of a jovial, beer-bellied fellow in a JOY OF SETS T-shirt. He stepped forward and filled his Styrofoam cup, half from one tank and half from the other. "I see," laughed one of his compatriots, "the game-theoretic solution!"

"I have another solution," said a thin man whose hands trembled slightly, apparently from a caffeine high. He filled his cup a quarter full from one of the tanks. "This is the asymptotic solution," he announced. "I'm already so wired I only want a little bit more; whatever is in the cup, I win."

A bystander who was a physicist rather than a mathematician didn't understand the need for all this sophisticated thinking. He pointed out that one tank was bigger than the other, and that the bigger tank presumably held caffeinated coffee since more people drink it than drink decaf. The mathematicians were taken aback by the simplicity of the solution. "Our different approaches," said the physicist, "remind me of an old joke. A physicist and a mathematician are flying cross-country together. Each is keeping a diary of the trip. They fly over a white horse in Iowa. The physicist writes, 'There is a white horse in Iowa.' The mathematician writes, 'There exists, somewhere in the Midwest, a horse, white on top.' "

∎

The search for mathematical truth first drew Erdős to Memphis in 1974. He came to visit two young men, Ralph Faudree and Dick Schelp, who had impressed him by solving a tricky problem in Ramsey theory. Erdős found their company sufficiently stimulating that he returned to the City on the Bluff every year, sometimes as many as four times a year, until 1996. So successful were these trips that he wrote forty-six papers with Faudree, and numerous others with Schelp and three other Memphis mathematicians, including the relocated Hungarian prodigy Béla Bollobás, who was fourteen, in 1957, when he won Hungary's infamous student math competition and was rewarded with an audience with Erdős at a Budapest hotel. "By the time I was nine, I very much wanted to do mathematics," recalled

Bollobás, "but I didn't know that there was such a thing as being a mathematician. I mean, why would anyone pay a person to be a mathematician?" Erdős evidently provided young Bollobás with a convincing answer, because their first meeting was the start of a forty-year collaboration that resulted in fifteen papers.

The last time Erdős and Bollobás saw each other in Memphis was March 1996. The annual Memphis conference coincided with Erdős's birthday, or as he always said, his "birthday wake." He turned eighty-three in 1996, and observed then that it was "better to be 38 than 83. Old age," he said, "is unpleasant in many languages." And then he repeated "Old age is unpleasant" in Hungarian (*"Rossz öregnek lenni"*), Hebrew, French, and German.

Old age or not, he hadn't lost any of his feistiness. He derided as "Fascist nonsense" the pending "monkey bill" that would restrict Tennessee teachers from teaching evolution. "You know what I say," Erdős said as he wrapped up his formal lecture. "Everything but mathematics must come to an end, and this also applies to lectures about mathematics. If I live, I will be back. If I don't, I won't."

Erdős, of course, didn't live, and his absence from the 1997 conference was conspicuous. His birthday, then, really was his wake. It fell to Béla Bollobás to kick off the first Paul Erdős Memorial Lecture. "This should be a joyous occasion," Bollobás told an audience of two hundred, "as we come together to celebrate the life of Paul Erdős." Not everyone looked joyous, though. Certainly not the keynote speaker whom Bollobás was introducing, Vera Sós, widow of Erdős's best friend and one of Erdős's most important collaborators in her own right, with thirty-five joint papers. "I first met Paul Erdős in 1948," she announced, "when one of his friends, my high school teacher Tibor Gallai, introduced us. The last time I met him was nearly forty years later in Warsaw. We had a long discussion the day before he passed away. He was working then on a paper." After recalling some of the problems he worked on, Sós put up copies of his letters on the Viewgraph. The audience was spellbound by Erdős's wiggly handwriting. "He would have been very pleased by a recent result," she said, "but did not live to see it." "Maybe it's in the Book," someone shouted, "and he's reading it right now." Everyone nodded in agreement. "If we want to remember Paul Erdős," Sós said, "we should prove some of his conjectures. There is no better tribute." Her talk was well received. Afterwards the audience was invited to flip through photo albums of Erdős and his mother.

"No son loved his mother more than Paul," said John Selfridge. "I got to know her well during the spring of 1966. She was a kind woman. We called her *Anyuka* [Mother] like Paul. She was there at the University of Illinois when Paul and I solved a 150-year-old problem," producing a proof that the product of consecutive integers is never a square, cube, or any higher power. The key ideas in the proof, Selfridge said, came from Erdős. "I was there working with him, improving a few of the things he did, putting my two cents' worth in. But I told him, 'This is really your solution. I only helped you.' He didn't exactly like that. He liked the idea that this was a joint paper, and so it was. I was certainly happy to be part of a very important paper. He was generous to a fault and inspired me to do things mathematically that I never thought I could do. I've been described as being a chronic nonpublisher. It took nine years for us to finish the paper, 'The Product of Consecutive Integers Is Never a Power,' and when it was finally published in 1975, it starts with a dedication to six of Paul's mathematical friends who had died from the time we began the work in 1966 to the time we actually got it ready for publication.

"When Paul died, I went to his funeral in Budapest. I hate funerals," said Selfridge, "but I'm glad I went." The official memorial service, at 11:00 A.M., on Friday, October 18, 1996, was one of the largest ever held in Hungary, with more than five hundred people in attendance, as if it were the funeral of a head of state. Then there was a mathematical memorial service that weekend, and finally on Monday, a few close friends escorted his ashes to his parents' grave in the Jewish cemetery in Rákoskeresztúr. "I waited," said Selfridge, for the others to leave, bent down, "and touched the twenty-five-year-old gravestone and said simply, 'Anyuka!' knowing how much she and Paul had loved each other. Paul and I loved each other too."

■

When the Memphis conference ended, Ralph Faudree, the chief organizer, invited everyone remaining over to his house for a "survivors' party." The Faudrees' dining-room table was draped with an old pink tablecloth covered with the Magic Marker signatures of attendees at previous years' parties. Erdős's shaky signature appeared several times. Next to the table was a peculiarly large black flag. "There's a story behind this," said Faudree. "Erdős had several birthday wakes here and said they should be celebrated by hanging out a black flag. Well, one year my wife Pat took him up on his whining and made him this black flag and a black cake." This year's survivors nodded familiarly at the flag and signed the tablecloth. Over pulled pork, baked beans, and beer, they exchanged Erdős stories.

"When he first came to Memphis in the seventies," said

Faudree, now a dean at the University of Memphis, "he only needed three hours of sleep. He'd get up early and write letters, mathematical letters. He'd sleep downstairs. The first time he stayed, the clock was set wrong. It said 7:00 but it was really 4:30 A.M. He thought we should be up working, so he turned on the TV full blast. Later, when he knew me better, he'd come up at some early hour and tap on the bedroom door. 'Ralph, do you exist?' The pace was grueling. He'd want to work from 8:00 A.M. until 1:30 A.M. Sure we'd break for short meals but we'd write on napkins and talk math the whole time. He'd stay a week or two and you'd collapse at the end. And after he left, you'd get constant phone calls from him for a few days. 'Have you finished the proof? Have you given it to the typist?' I told people that I once tried sleeping downstairs in what we called the Erdős suite. It didn't help my mathematical stamina, but I invited other mathematicians to try."

"Faudree's failure to keep up with Erdős was hardly unique," said Michael Jacobson. "I met Erdős in 1976, when I was a graduate student at Emory. He gave a lecture and declared at the end, 'My brain is open.' It was an invitation to talk mathematics with him. So I went up and told him the problem I was working on in combinatorial set theory. He responded, without a moment's hesitation, 'I think you need to find another problem.' In other words, he was saying the problem was too difficult. I was devastated—I was just a student. At least I wasn't so upset that I quit mathematics. Seven years later, when I was at Louisville, I reminded him of our first meeting and the devastating thing he had said. But he didn't remember the conversation, and asked me to state the problem again. I told him, and he replied, 'Well, it was too hard. It still hasn't been solved.' He was absolutely right. His mathematical insight was amazing. Today, twenty-one years later, the problem still hasn't been solved.

"When I met him in 1976, he was always short of time. He'd size people up quickly. He was on the lookout for people who really had a spark of genius. When he dismissed my problem as too hard, I guess he decided I didn't have that spark. Although friends point out that he must have thought I had something going because later, for more than a decade, he came to Louisville once a year and stayed a week each time. It was 1983 when he came for the first time. I was single then, and he stayed with me. It was very exciting. I had heard stories, of course, of how hard he worked—and I looked forward to working hard—but I wasn't prepared for just how hard. That first day we did mathematics until one in the morning. I was completely drained. I went upstairs to bed, and he stayed downstairs in the guest room. At 4:30 A.M. I heard pots banging in the kitchen. He kept banging them. It was his way of telling me to get up. I stumbled downstairs about six. What were the first words out of his mouth? Not 'Good morning' or 'How'd you sleep?' but 'Let $n$ be an integer. Suppose $k$ is . . . ' I was half-naked, with just a bathrobe on and my eyes blurry and partially shut. I drew the line there. I told him I couldn't do mathematics before I took a shower."

"Even though Erdős only got three hours of sleep a night," added Ronald Gould, a mathematician at Emory, "he did take little naps throughout the day. But he was doing math even then. I once spent an evening explaining a proof to him. He'd doze off, and I'd stop speaking because I thought he wasn't paying attention. As soon as I stopped, though, he'd raise his head and tell me to continue. This went on all evening. He'd doze, wake up if I paused, doze, wake up. And you know what was astonishing? After all that, he understood the proof!"

It wasn't only the sheer hours he put in doing mathematics that was unusual, but also the occasions he decided to do it. A funeral was as good a place to do mathematics as a wedding was, and holidays didn't stand in the way either. "In 1983 my family was in Budapest on Christmas Eve," said Mike Plummer of Vanderbilt University. "Christmas is a very big deal there. Families get together and open presents. The city stops. Everything closes. The buses don't run. You have to book a taxi days ahead of time. We were all dressed up, waiting for a taxi, to go see some friends. We were standing around when all of a sudden there is a pounding on the door. I opened it and there was Erdős. 'Plummer,' he said, 'Merry Christmas. Let $f$ of $n$ be the following function . . .' So we talked math until the cab showed up. It was hard to shut him up and take off."

Faudree, like the other Uncle Paul sitters, did not find him an easy houseguest: "He was very sensitive to temperature and circulation. Our house has two stories, with an AC control for each floor. The downstairs control was outside the room he usually slept in. One time another visiting mathematician was staying in that room, so we put Erdős upstairs. He was hot so he went downstairs and turned the AC up. That did nothing to the temperature upstairs, so he kept going downstairs and cranking the AC up. When we got up in the morning, there was frost on the downstairs windows.

"He didn't like closed rooms. So another time we left the downstairs window open for him and went up to bed. Later there was a torrential rainstorm. He came upstairs, woke us up, and announced, 'It's raining in the window. You'd better do something.' "

Although he was hopeless with material things, he was always kind to people. "Whenever he came back to Hungary after being away," said Vera Sós, "he'd call the old ladies. He always spent the first two or three days visiting the mothers and widows of mathematicians."

In 1994 in Budapest, Erdős took Zoltán Füredi, a young Hungarian combinatorialist who had collaborated with him on nine papers, to see an old friend, the mother-in-law of a mathematical colleague. "She was in a wheelchair," recalled Füredi, "and for some reason she was very bitter. She was complaining constantly to her maid: 'Don't put that there' and 'Why don't you serve Herr Professor Erdős first?' and other stupid things. Then we all talked—Erdős did most of the talking—for about an hour. As we left and the maid escorted us out, he gave her a very generous tip, as if to acknowledge that she had been treated unfairly. I was impressed. He was always able to be independent, and he never forgot to do the kind thing.

"He really was an extraordinary person. I was going through an ugly divorce the past three years. He'd call every two weeks. Each time he'd ask about the kids and assure me he felt sorry for them. Then he'd feed me these incredibly intriguing mathematical problems. He was trying to help in his own way by keeping me going mathematically. I've been struggling to finish a paper we started in 1992. The fact that he died is forcing me now to work because I owe it to him to finish the paper soon."

"He certainly was a caring person," said Faudree. "He cared about my kids, always asked about them, and worried about them if anything was wrong. When I went to stay in Hungary for the first extended time, in 1981, he told everyone to take care of me and my family. He had one person worry about our medical care, another person worry about our kids' schooling. He paid attention to people in need. When Russian mathematicians arrived penniless in Hungary, he gave them every cent he had.

"He also cared about things you cared about, even things he really didn't like. For instance, he didn't like dogs. Whenever a dog approached him, he'd say, 'Is that Fascist hound going to bite me?' But we had a couple of dogs, and he tolerated their greeting him by jumping on him. 'Good,' he'd say, when they leaped on him, 'they remember me.' We used to leave coffee and cereal out for him because he'd get up earlier than us. One day when I came down to the kitchen there was cereal, lots of cereal, all over the floor. I didn't understand how it got there. Even if he opened a new box and had to struggle to rip the plastic, that much cereal couldn't have shot out. I couldn't figure it out, so I just swept it up. The next morning I came down and there was cereal all over the floor again. Erdős was sitting there, dropping fistfuls of cereal, trying to feed the dogs."

Erdős was hard on everyone's floors. "Paul liked to wash his hands all the time," recalled Dick Schelp. "He really had this thing about picking up germs. You'd leave out a fresh towel but he wouldn't use it. He'd just shake the water off. My son said that after he's been in the bathroom, it's like walking in the middle of a swimming pool. There's water all over."

At least water can be cleaned up. Some Uncle Paul sitters, though, were not as lucky. "He was staying in one of our kids' bedrooms," said Plummer. "He had various skin afflictions and brought with him different wet and dry substances, different lotions and powders, to put on himself depending upon how his skin felt. He had taken a bath and his skin bothered him. He wasn't sure which substances would help, so he had lathered himself with lotion and talc. He managed to spill the lotion and talc all over the floor. He tracked footprints across the bedroom. The prints are still there!"

Erdős's caretakers drew some comfort from the fact that Erdős was as hard on his own belongings as he was on other people's. "I remember him ruining a brand-new suit," said Faudree, "just after someone gave it to him. We were driving from Memphis to Louisiana State. Normally we took Interstate 55 but Erdős insisted on taking a more scenic route. So we decided to go through Natchez and visit some antebellum homes. We were walking along the Mississippi behind one of these homes. At some point we had gotten behind a chain-link fence with little spikes on the top. Eventually the fence met the river and we couldn't go any further. Paul, always impatient, insisted we scale the fence. Schelp and I easily made it over. Then Paul climbed up but sat right down on the spikes. As he jumped off, we heard a loud ripping sound. He had a foot-long tear in the seat of his pants. He was very upset. I said, 'Paul, do you have another pair?' 'No, no,' he said, 'this is the only good pair. I have one other pair but they bother my skin.' When we got to Louisiana, he went straight to his room. We were going to send the pants to a tailor. But he came out with a smile on his face and said he'd looked in the mirror and saw that the suit was okay because the coat covered the rip. The next day we were walking to lunch. And, impatient again, he started to head down an alley he thought was a shortcut. I said, 'I don't think you can get through that way.' And he said, 'Why not?' I said, 'I think there's a fence at the end.' He smiled and said, 'There is an old Hungarian saying: Do not mention rope in the house of a man whose son has just been hanged.' "

# "WE MATHEMATICIANS ARE ALL A LITTLE BIT CRAZY"

I am not a mathematician. I have never proved a theorem, let alone offered a surprising conjecture. My Erdős number is infinity. Yet there I was at the Memphis survivors' party, feeling at home, because I too had been touched by Paul Erdős—his innocence, his frailness, his magnificent obsession, his generosity, even his impish take on God.

I looked around the party at his followers. They were a playful and diverse bunch. They came in all shapes and sizes. They were round and frail, slobs and clotheshounds, men and women, oldsters and adolescents, womanizers and wallflowers—at least a dozen nationalities were represented—all united in the common quest for universal mathematical truth. The Church of Mathematics has a big-tent philosophy, welcoming anyone who is serious about the quest. Many of the Memphis survivors were smart and witty, spouting aphorisms that sounded like Erdős. A few were shy and sat alone, quietly nursing beers. Most seemed perfectly sane. One or two had the nervous look of outpatients trying to keep a lid on things.

In mathematics, madness often comes with the territory. "When I first met Landau in 1935 in Cambridge," Erdős liked to recall, "he told me, '*Wir Mathematiker sind alle ein biβchen meschugge*,' which means, We mathematicians are all a little bit crazy. And in 1932, I met a Hungarian mathematician called Sidon, who worked mostly in trigonometric series. He was a very good mathematician, but he was a bit crazier than the average mathematician. In fact, he was a borderline schizophrenic. They tell about him that he usually talked this way to you," he said, imitating Sidon, by turning away and leaning close into the wall. "But when he talked about mathematics, he talked sense. . . . In 1937 when Turán and I visited him—he also had a persecution complex—he opened the door a crack and said, 'Please come rather at another time and to another person.' *Kérem jöjjenek máskor és különösen máshoz*— that sounds better in Hungarian. But later on he was again reasonable. Now anyway, let us forget this sad story. Actually, he had a curious death. He died like Cyrano de Bergerac. A ladder fell on him and broke his leg, and he died in the hospital of pneumonia. But anyway, I am sorry that he didn't live to see how popular he became. Every mathematician seems to know him now."

"In mathematics you can be pretty far away from the mean," said Graham, "and still survive—that is, still earn a living. In mathematics you see a lot of people who couldn't be, say, salesmen. But they have enough control to do math. Some people are beyond that, like John Nash, for example." Nash was a mathematician at Princeton who won a Nobel Prize in economics in 1994 for a twenty-seven-page doctoral dissertation on game theory written almost half a century earlier, in 1949, when he was twenty-one. His professors called him "young Gauss" because of his lightning speed in solving problems, and *Fortune* magazine, in 1958, hailed him as the brilliant star of the "new mathematics." But then, at the height of his creativity, schizophrenia seized him. He heard voices, believed strangers were spying on him, and drew inexplicable connections between people and numbers. He shared the connections in one-line postcards: "I rode on bus No. 77 today and it reminded me of you." Soon he was committed, at the age of thirty, to a psychiatric hospital in Belmont, Massachusetts, alongside the poet Robert Lowell. Nash wasn't cured. He ended up wandering the halls of Princeton, hanging out in the campus library, and becoming a character, the Phantom of Fine Hall, in Rebecca Goldstein's novel *The Mind-Body Problem*. More than twenty years later, Nash, without drugs or treatment, had a spontaneous remission, and was well enough to go to Stockholm, when the Nobel Committee, after heated deliberations, decided it was appropriate to give the prize to someone who was mentally ill.

Nash and Sidon were not isolated cases. Cantor had become delusional, Gödel had grown paranoid, and the mathematician who wouldn't perform on composite-numbered days had turned criminally violent. Then there is the Unabomber, Theodore Kaczynski, who received a Ph.D. in mathematics from the University of Michigan in 1962.

"I have a theory," said Graham about mathematics and madness. "In so many areas of mathematics it seems natural or appropriate to create your own mathematical world. You have a lot of choices. I want to consider structures that have thus-and-such properties. I want this structure and not that one. Physicists don't have that freedom: they are confined by our actual world and try to study it. In physics, you can't suppose that gravity is inverse-cubed when in fact it is inverse-squared. In mathematics, you do things like that all the time. Remember what Straus said about Einstein: that he was afraid of going into mathematics because it wasn't clear what the good questions were. In physics, it is clear what the important questions are and you go address them. In mathematics, how do you even know where to start?

"There's a premium in mathematics on doing something new, something different. It can carry over to the rest of your life. There's a tendency to make up your own rules. $A + B$ is not equal to $B + A$. Well, everyone knows it is, but suppose it isn't. Now what have you got? Why should you drive on the right when you can drive on the left?

"Of course there are also people who really aren't all that strange who deliberately act in outlandish ways because that's how geniuses are supposed to act. Einstein once complained to a friend that the price of scientific fame was people coming up to him all the time wanting to talk. You could end that in a moment, the friend said, by cutting your hair.

"There are problems in graph theory called *extremal problems*. They are problems about extremes, like 'What's the largest number of edges a graph can have?' Erdős was into questions like that. I had heard about a European aristocrat who bought out an opera house and gave tickets to his friends and put all the bald ones in a certain position so that when he looked down from the balcony their bald heads spelled out something spectacular, unbeknownst to them. That gave me the idea of holding a conference of extremal mathematicians. I'd invite all the strangest people I knew over at the same time, not tell them why they were invited, and see if they could figure it out."

Many people who go into mathematics are seeking refuge from the rest of the world. "I believe with Schopenhauer," said Einstein, "that one of the strongest motives that lead men to art and science is to escape from everyday life with its painful crudity and hopeless dreariness, from the fetters of one's own ever-shifting desires. A finely tempered nature longs to escape from the personal life into the world of objective perception and thought." Bertrand Russell had a troubled adolescence. He contemplated suicide

but didn't go through with it because there were mathematical problems left to solve. Others have taken their own
lives because they couldn't solve the very problems that
kept Russell going. Logicians joke about a black theorem,
which if you discover it, drives you mad. Short of insanity,
said Graham, "you see a lot of people, even quite good
people, bail out. They get depressed when they recognize
the impossibility of knowing it all. If you're a stamp col-
lector, you can collect all the stamps. In mathematics, you
can't prove all the theorems. It can be very discouraging."

■

Ron Graham introduced me to Erdős in 1986. At our first
meeting Erdős barely noticed me, head deep that he was
in one conjecture or another. I decided then that I was
going to follow him for a few weeks as he traveled around
and showed up unannounced on his colleagues' doorsteps,
eager to skip the small talk and plunge right into mathematics. Ron called ahead so that people Erdős might decide
to visit would let me tag along. I slept where he slept and
stayed up nineteen hours a day, watching him prove and
conjecture. I felt silly not being able, at the age of thirty,
to keep up with a sickly looking seventy-three-year-old
man. I suppose I could have shared his pills, but the only
stimulant I took was caffeine.

I kept a diary of my travels with Erdős and wrote up
his story in *The Atlantic*. When the piece was published,
it got a lot of attention. A few years later, I ran into Erdős
and asked what he thought. "Of what?" he asked. "Of the
story I wrote," I said. "Ah! The story," he said. "Do you
have it?" I gave him a copy and he held it up close to his
face. Squinting, he studied it, page for page, for what
seemed like eternity. "What do you think?" I finally

asked. He shook his head from side to side. "It's okay," he said. "Except for one thing." I thought he was about to correct my explanation of Chebyshev's discovery or the Four Color Map Problem or the Prime Number Theorem. "You shouldn't have mentioned the stuff about Benzedrine," he said. "It's not that you got it wrong. It's just that I don't want kids who are thinking about going into mathematics to think that they have to take drugs to succeed." That was Erdős, always thinking about the epsilons.

# ACKNOWLEDGMENTS AND
# SOURCE NOTES

---

This book is in large part a work in oral history based on the recollections of Erdős, his collaborators, and their spouses. *The Man Who Loved Only Numbers* could not have been written without the kind help of Ronald Graham, who put me in touch with Erdős's coworkers, shared hundreds of letters and e-mails about him, and spent many days helping me understand the mathematics. Those who want to know more about Erdős's contributions to mathematics should read Graham's latest book, written with Fan Chung, *Erdős on Graphs: His Legacy of Unsolved Problems*. Graham also provided the Erdős letters for the endpapers and unattributed photographs. I am grateful to Andrew Vázsonyi for telling me especially about the early years in Budapest. Thanks, also, to Fan Chung, Anne Davenport, Pat Faudree, Ralph Faudree, Peter Fishburn, Magda Fredro, Zoltán Füredi, Ronald Gould, Richard Guy, Melvin Henriksen, Michael Jacobson, Carole Lacampagne, Aaron Meyerowitz, Melvyn Nathanson, Chip Ordman, Mike Plummer, George Purdy, Bruce Rothschild, Dick Schelp, John Selfridge, Alexander Soifer, Vera Sós, Joel Spencer, Maryann Spencer, Louise Straus, Françoise Ulam, Laura Vázsonyi, David Williamson, Herb Wilf, and Peter Winkler for sharing stories about Uncle Paul. Thanks to editors at *The Atlantic*, William Whitworth and C. Michael Curtis, for giving me the assignment to profile Paul Erdős (November 1987) and providing the seed money for me to follow him around. Thanks to Bob Miller, Rick Kot, and Will Schwalbe for helping me turn Erdős's story into this book.

I have only given sources for written Erdős quotations. If the source of a quotation is not shown below, it is because Erdős made the remark directly to me or it was told to me by someone I interviewed.

## CHAPTER 0

p. 6 "There is an old saying." Quoted in Alexanderson [December 1979], 87.

7 Kurdish limerick. Quoted in ibid., 91.

8 "When we met." Quoted in Hargittai [Fall 1997], 38.

10 "In the early 1960s, when I." Larman [September 28, 1996].

11 "He was the Bob Hope." Nathanson [October 19, 1996].

12 "He still did very good things." Quoted in Alexanderson [December 1979], 85.

13 "I was told several years ago." Goffman [September 1969], 791.

14 "One evening in the seventies." Rota [November 1996].

14 The url of Jerrold Grossman's Erdős Number Project is *http://www.oakland.edu/~grossman/erdoshp.html*

21 "Once I spent a few days with Paul." Pach [Spring 1997], 41.

## CHAPTER 1

29 "Mathematics is the surest way." Quoted in Saville [March 10, 1995].

29–30 "When the grown-ups went for a walk." Pach [Spring 1997], 45.

30 "It is most important in creative science." Ulam, Stanislaw [1976], 55.

30–31 "A chess problem." Hardy [1940b], 80, 89.

31 "no chess problem has ever affected." Ibid., 90.

32 Reverend Hugh Jones and the octary system. Kramer [1970], 13.

32 "much more direct contact." Hardy [1940b], 128.

33 The url of the largest known prime is *http://www.utm.edu/research/primes/largest.html* and the url of the Great Internet Mersenne Prime Search is *http://www.mersenne.org/prime.htm*

34 Frank Nelson Cole and $2^{67} - 1$. Quoted in Gardner [March 1964], 126.

36–37 "Between interrogations, always blindfolded." Cooper, Roger [April 13, 1991], 236, 237.

37–38 "He half-dozed through." Kac [1985], 90.

38–39   Hal Shapiro and the upstaged paper. Bellman [1984], 114, 115.

40   Gauss's letter to Johann Franz Encke, 1849. Quoted in Hall [1970], 14, 15.

41–42   "Erdős contributed an enormous." Guy [January 10, 1997].

42   "He had the unique ability." Quoted in Roddy [October 18, 1996].

43   "there is a very high degree." Hardy [1940b], 113.

47–48   Williamson and Erdős. Conversation with Williamson; and quoted in Babai [January 1998b], 66, 67.

50   "whom Erdős could beat only rarely." Golomb [Fall 1996], 7.


CHAPTER 2


60–61   "Give me a four-digit number." Shoe store dialogue from conversation with Vázsonyi; and from Vázsonyi [December 1996], 59, and [March 1997], 19.

62   "thousands of people streamed." Lukacs [1988], 12.

63   "illicit trading, adultery, puns." Hungarian journalist quoted in Rhodes [1986], 105.

64   "has never been an agreeable." Krúdy quoted in translation in Lukacs [1988], 21, 22.

65   Yom Kippur and Maupassant. Babai [1996], 52.

65   1910 statistics on Jewish occupations. Rhodes [1986], 105.

67   "The rousing tunes." Koestler [1952], 63.

67–68   "The people of Budapest." Ibid., 65, 66.

68   Edward and Ilona Teller's ordeal during Kun's regime. Blumberg and Owens [1976], 17–19.

69   "then the mecca of scientists." Ibid., 20.

70   "There were a lot of anti-Semitic acts." Quoted in Tierney [October 1984], 46.

70   "Apuka!" Quoted in Babai [1996], 56.

72   "Women are bosses and men slaves." Quoted in Alexanderson [December 1979], 90.

73   FBU. Henriksen [1997].

73   Barbara Piranian and Erdős. Quoted in ibid.

73   "complained of the phone ringing." Smith [1997], 83.

74   "Paul, then still a young student." Szekeres [1973], xix, xx.

76 "We soon realized." Ibid., xxi.

78 "From 1934 he hardly ever slept." Bollobás [December 1996], 4.

78 "he had intended." Bollobás [October 17, 1996], 584.

79 "could discharge certain duties." Newman [1956], 2025.

79 "profound conviction that the truths." A.H.S. [January 22, 1948], 210.

80 Russell and Littlewood's praise of Hardy's good looks. Quoted in Bollobás [1986], 120.

80 "strikingly handsome and graceful." Newman [1956], 2025.

80 "a kind of Red Indian bronze." Snow [1966], 21.

80 "a maximal theorem." Quoted in Newman [1956], 2024, 2025.

80 "Bradman is a whole class." 1938 Hardy postcard quoted in Snow [1966], 37.

80 "must not have been very helpful." Newman [1956], 2026.

81 "even the womanisers." Deacon [1985], 139.

81 "a nonpracticing homosexual." Quoted in Kanigel [1991], 139.

81 "In his fifties he could usually." Snow [1966], 28.

81 "(1) prove the Riemann Hypothesis." Quoted in Nature [May 22, 1948], 798.

82 "It is a melancholy experience." Hardy [1940b], 61.

82 "Galois died at twenty-one." Ibid., 71, 72.

82–83 "The discovery of Ramanujan." Quoted in Alladi [December 19, 1987], 8.

83–84 Ramanujan's letter to Hardy. Quoted in Berndt and Rankin [1995], 21, 22.

85 "that if he could find a proof." Quoted in Russell [1956], 19.

85 "An equation for me." Quoted in Kanigel [1991], 7.

85 "I owe more to him." Hardy [1940a], 2.

86 "What happened next." Kanigel [1991], 294.

86 "Although Hardy was modest." Quoted in Alladi [December 19, 1987], 8.

86 "could remember the idiosyncrasies." Hardy [1940a], 12.

87 "pinnayopp-play oopshiday dovn tsockay." Quoted in Smith [1997], 82.

87 Table of number of primes up to an integer *n*. Regis [1987], 74.

88–89 Highly composite numbers. Honsberger [1985], 193, 194.

93 "was doing research on." Vázsonyi [December 1996], 59.

**CHAPTER** *e*

96    "twenty-seven years old, homesick." Ulam, Stanislaw [1976], 134, 135.

97    "that he was a pure mathematician." Frisch [1979], 79.

98    "He was always strongly." Bellman [1984], 102, 103.

99    "It was a harmless situation." Quoted in transcript of outtakes of Csicsery [1993].

100    "I was thinking." Quoted in Babai [1996], 63.

100    "By the evening, they cleared." Quoted in transcript of outtakes of Csicsery [1993].

100    "was no longer entirely penniless." Ulam, Stanislaw [1976], 135.

101    "While he meant much to our group." Golomb [Fall 1996], 6.

102    "What's your profession?" Pach [Spring 1997], 40.

103    "Under orders from Stalin." Babai [1996], 65, 66.

103    Jay-walking incident. Conversation with Rothschild; and Babai [1996], 79.

103–104    "I remembered suddenly Plato's." Ulam, Stanislaw [1976], 175–78.

104–105    "I agreed." Ibid., 179.

105    "Ulam was always afraid." Erdős [1985], 447, 449.

105    "a long and very boring paper." Quoted in Gardner [March 1964], 122.

106–107    "Ulam's doodles in the twilight zone." Ibid., 124.

109    "without the risk of suicidal war." Ulam, Stanislaw [1976], xiii.

110    "a horrible dream." Quoted in Hardy [1940b], 83.

111    "Gödel I talked with a great deal." Quoted in Alexanderson [December 1979], 89.

112–113    "Originally viewed as both a tool." Osserman [1995], 6.

114–115    "If God exists." Dostoyevsky [1880], 270, 271.

115    "A scientist can hardly meet." Russell's letter to Frege quoted in Davis and Hersh [1980], 332.

115–116    "an affinity with the ancient Greek." Russell [1951], 221, 222.

116    "that a class sometimes is." Russell [1959], 75.

116–117    "What we have experienced." Quoted in Regis [1987], 53.

117    "God exists since mathematics." Quoted in Singh [1997], 160.

118    "I wanted certainty." Russell [1956], 53.

120    "I call our world Flatland." Abbott [1884], 1.

120–121    "however successful the theory." Eddington [1920], 131.

121–122    Gödel and the citizenship exam. Casti [1990], 373, 374.

122    "We have to divide up our time." Quoted in Regis [1987], 41.

122–123    "I think that the atomic bomb." Quoted in Arkin [April 1976], 3.

126    "Einstein often told me." Straus [1983], 245, 246.

127    "The only thing that bothered me." Quoted in Henriksen [1997].

127    "We said, 'Paul'." Quoted in ibid.

128    "No reason was ever given." Ibid.

130    "The central theme of their 'compositions'." Pach [Spring 1997] 44.

## CHAPTER 3

132–135    "When I returned from the United States." Erdős [March 1971], 3–12.

135    "pouring problems on us." Pelikán [1996], 17.

135    Natalija story. Ivić [1997].

136    "When he asked a father of four." Pelikán [1996], 17.

136    "We did not see him." Golomb [Fall 1996], 7.

137    "His requests were rejected." Golomb [May 27, 1977], 938.

140    "It's a very complicated situation." Quoted in Tierney [October 1984], 46.

141    "early one Sunday morning." Vázsonyi, unedited version [December 1996]. Conversation with Vázsonyi resumes with "He told me he was at UCLA."

## CHAPTER π

152    "and on the first day things." Quoted in Albers [November 1996], 20.

156 Greedy expansion of 5/121. Michael N. Bleicher, *Excursions into Mathematics* (Worth Publishers, 1969). Quoted in Gardner [October, 1978], 26.

162 "I have never done anything 'useful'" Hardy [1940b], 150.

162–163 "This is the remarkable paradox." Tierney [October 1984], 43.

163 "everything that can be used." Quoted in Arkin [April 1976], 4.

163 Vázsonyi contemplates giving up mathematics: In conversation; and in Vázsonyi [December, 1996], 59.

163 "I am praying for your soul." Quoted in Kac [1985], 93.

171 "A plumber must cut a set of pipes." Graham [March 1978], 130, 131.

171 "a person confronting a standard." Graham [1978], 203.

**CHAPTER 4**

180 "questions like 'When do you.'" Nelson, Penny, Pomerance [Spring 1974], 87.

181 "He was a gentle man." Quoted in Petrillo [January 11, 1997], B-3.

181 "And thus Hank Aaron." Quoted in Mackenzie [February 7, 1997], 759.

183–184 "This is probably the only case." Quoted in Singh [1997], 229.

184 "Who's Barbara Walters?" Quoted in Dietrich [August 14, 1996], 1.

184–185 Fake e-mail news story. Quoted in Stewart [1987], 46.

190 To find out the latest on the largest Sophie Germain primes, check out the url *http://www.utm.edu/research/primes/largest.html#Sophie*.

190 "Unfortunately, the depth of my intellect." Quoted in Dalméldico [December 1991], 119.

191 "A taste for the abstract sciences." Quoted in Bell [1937], 262.

192 "believed that a mathematician." Ibid.

193 The story of Paul Wolfskehl's aborted suicide comes from Davis and Chinn [1969]. In a chapter called "The Problem That Saved a Man's Life," Davis and Chinn attribute the details of the story to "the renowned mathematician Alexander Os-

trowski. Professor Ostrowski himself heard the story many years ago and claims there is more to it than mere legend." But in "Paul Wolfskehl and the Wolfskehl Prize" (*Notices of the American Mathematical Society*, November 1997, vol. 44, no. 10), Klaus Barner suggests an alternative explanation. Wolfskehl, who suffered from multiple sclerosis at an early age, was forced by his family to marry an older spinster, Marie Frolich. "His wife, Marie, revealed herself as an evil Xanthippe, who made the last years of his life a living hell." He set up the prize, Barner believes, because "number theory, the only true love in his life, had given his last years some meaning. Perhaps the desire not to leave his entire fortune to his loveless wife, Marie, also had a part to play?"

194    "at the moment." Quoted in Ribenboim [1979], 15, 16.

195    "have experienced something." Bell [1937].

195    "one-tenth genius." Ibid., 20.

195    "In spite of his lofty thoughts." Ibid., 42.

196    "looked so simple." Quoted in Singh [1997], 6.

196    "it is an interesting question." Ribet and Hayes [May–April, 1994], 144.

197    "Of course the Taniyama-Shimura." Quoted in Singh [1997], 226.

197    "For the first few years." Quoted in Aczel [1996], 118.

197    "entering a darkened mansion." Quoted in Cipra [1996], 3.

198    "There was only one possible climax." Quoted in Aczel [1996], 3.

198    "I came relatively early." Quoted in Singh [1997], 270.

198    "This proves Fermat's Last Theorem." Quoted in Aczel [1996], 4.

198    "All number theorists." Quoted in Kolata [June 29, 1993], C11.

198    only one tenth of 1 percent. Kolata [June 24, 1993], D22.

199    "Having solved this problem." Quoted in Singh and Ribet [November 1977], 73.

199    "If I've helped to counter." Quoted in Gleick [October 3, 1993], 53.

## CHAPTER 5

204    "The audience gasped." Ulam, Stanislaw [1976], 288.

205    Smith's phone number. Wilansky [1982], 21.

206 "be obtained by showing." Oltikar and Wayland [January 1983], 37.

207 "a squalid relic of the Middle Ages." Bell [1937], 221.

212–213 "A cow from a cow." Gardner [June 1977], 131.

216 "The nine Indian figures." Quoted in Gies [1969], 58.

218 "The divergent series." Quoted in Maor [1987], 33. The limits given for the harmonic series also come from Maor.

221 "I see it but I don't believe it." Quoted in Stewart [1987], 67.

222 "From the paradise." Quoted in Gardner [March 1966], 112.

223 "I entertain no doubts." Quoted in Dauben [1979], 147.

224 "From me Christian Philosophy." Quoted in ibid., 147.

224 "which will not fail to terrify." Quoted in ibid., 282.

224–225 "probably the greatest." Quoted in Gardner [March 1966], 112.

225 "God made the integers." Quoted in Stewart [1987], 67.

225 "a disease from which." Quoted in ibid., 67.

225–226 "Mathematicans knew there was only." Barrow [1992], 215.

226–227 "If I were alive in a thousand years" through "you would have otherwise." Erdős and Graham exchange about the Continuum Hypothesis. Reprinted with the permission of the Canadian Broadcasting Corporation. It is an excerpt of the outtakes from the radio program "Math and Aftermath: A Tribute to Donald Coxeter," broadcast on the series *Ideas* on May 13, 1997. The program featured *Ideas* host Lister Sinclair and was produced by Sara Wolch.

228 "I shall risk nothing." Quoted in Bell [1937], 464.

230 "Two people walking in opposite directions." Peterson [1998], 40.

**CHAPTER 6**

234 "You pick door No. 1." Vos Savant [September 9, 1990], 15.

235 Letters from Robert Sachs and Scott Smith. Vos Savant [December 2, 1990], 25.

235 "when you switch, you win two." Ibid.

236 E. Ray Bobo letter. Vos Savant [February 17, 1991], 12.

236 "When reality clashes." Ibid.

238 "After spending a lot of time." Quoted in Eckhardt [1987], 131.

241 "I should be put under tutelage." Quoted in Alexanderson [December 1979], 88.

244 "probably a square for the very last time." Quoted in Cerabino [March 11, 1994], 1B.

### CHAPTER 7

252–253 "By the time I was nine." Quoted in Marshburn [Summer 1996], 7.

253 "Old age is unpleasant." Quoted in ibid., 7.

253 "fascist nonsense." Quoted in Beifuss [March 25, 1996], 1B.

253 "You know what I say." Quoted in Marshburn [Summer 1996], 7.

### CHAPTER ∞

263–264 "When I first met Landau." A story he told repeatedly in lectures. Quoted, for instance, in Csicsery [1993].

264–265 Details of Nash's life. Nasar [November 13, 1994].

266 "I believe with Schopenhauer." Quoted in Regis [1987], 43.

# BIBLIOGRAPHY

A.H.S., January 22, 1948. "Obituary: G. H. Hardy," *The Oxford Magazine*. vol. LXVI, no. 9.

Abbott, Edwin A. 1884. *Flatland: A Romance in Many Dimensions*. HarperCollins, 1983.

Aczel, Amir D. 1996. *Fermat's Last Theorem: Unlocking the Secret of an Ancient Mathematical Puzzle*. Four Walls Eight Windows.

Albers, Donald. September 1995. "Making Connections," *Math Horizons*. Mathematical Association of America.

————. November 1996. "A Nice Genius," *Math Horizons*. Mathematical Association of America.

————, and Gerald L. Alexanderson, eds. 1985. *Mathematical People: Profiles and Interviews*. Birkhäuser.

Alexanderson, Gerald L. December 1979. "Interview with Paul Erdős." In Albers and Alexanderson, eds., 1985.

Alladi, Krishnaswami. December 19, 1987. "Ramanujan—An Estimation," *The Hindu* [Madras, India].

————. December 26, 1996. "Erdős and Ramanujan," *The Hindu* [Madras, India].

Allstetter, Billy. Spring 1990. "How Many Mathematicians Does It Take to Be a Professor, Research Director, Journal Editor, and World-Class Juggler? Just One—Ronald Graham," *Rutgers*, vol. 69, no. 1.

Andor, Faragó. 1927. *Középiskolai Matematikai Fizikai Lapok*. Budapest.

Arkin, Joseph. April 1976. A taped interview with Paul Erdős.

Babai, László. 1996. "In and Out of Hungary: Paul Erdős, His Friends, and Times." In *Combinatorics, Paul Erdős Is Eighty*, vol. 2. Bolyai Society Mathematical Studies, Budapest.

————. January/February 1997. "Paul Erdős and the Theory of Computing," *SIAM News*.

————. January 1998a. "Finite and Transfinite Combinatorics," *Notices of the American Mathematical Society*, vol. 45, no. 1.

————. January 1998b. "Paul Erdős Just Left Town," *Notices of the American Mathematical Society*, vol. 45, no. 1.

Barner, Klaus. November 1997. "Paul Wolfskehl and the Wolfskehl Prize," *Notices of the American Mathematical Society*, vol. 44, no. 10.

Barrow, John D. 1992. *Pi in the Sky: Counting, Thinking, and Being*. Little, Brown.

Beckmann, Petr. 1971. *A History of π (PI)*. St. Martin's Press, 1974.

Beifuss, John. March 23, 1996. "Minister of Math, Calculating Genius at Home with Peers," *The Commercial Appeal* [Memphis].

————. March 25, 1996. "Give Me a Break," *The Commercial Appeal* [Memphis].

Bell, E. T. 1937. *Men of Mathematics*. Simon & Schuster, 1986.

Bellman, Richard. 1984. *Eye of the Hurricane: An Autobiography*. World Scientific.

Berndt, Bruce C., and Robert A. Rankin. 1995. "Ramanujan Letters and Commentary," American Mathematical Society.

Bernstein, Jeremy. 1984. *Three Degrees Above Zero: Bell Labs in the Information Age*. Scribners.

Blumberg, Stanley A., and Gwinn Owens. 1976. *Energy and Conflict: The Life and Times of Edward Teller*. G. P. Putnam's Sons.

Bollobás, Béla. 1986. *Littlewood's Miscellany*. Cambridge University Press.

————. October 17, 1996. "Paul Erdős (1913–96)," Obituary, *Nature*, vol. 383, p. 584.

————. December 1996. "A Life of Mathematics—Paul Erdős, 1913–1996," *Focus*, vol. 16, no. 6. Mathematical Association of America.

————. 1997. "Paul Erdős—Life and Work." In Ronald L. Graham and Jaroslav Nešetřil, eds. 1997a.

Borwein, Jonathan M., and Peter B. Borwein. February 1988. "Ramanujan and Pi," *Scientific American*.

Bowers, John F. December 22/29, 1983. "Why Are Mathematicians Eccentric?" *New Scientist*.

Casti, John L. 1990. *Searching for Certainty*. William Morrow.

Cerabino, Frank. March 11, 1994. "Mathematicians Hit Critical Mass at FAU Gathering," *Palm Beach Post* [Florida].

Chung, Fan, and Ron Graham. 1998. *Erdős on Graphs: His Legacy of Unsolved Problems*. A K Peters.

Cipra, Barry. 1993. "Number Theorists Uncover a Slew of Prime Impostors." *What's Happening in the Mathematical Sciences*, vol. 1. American Mathematical Society.

———. 1994. "A Truly Remarkable Proof." *What's Happening in the Mathematical Sciences*, vol. 2. American Mathematical Society.

———. February 4, 1994. "Math Attendees Find There's Life After Fermat Proof," *Science*.

———. 1996. "Fermat's Theorem—At Last!" *1995–1996 What's Happening in the Mathematical Sciences*. American Mathematical Society.

Clawson, Calvin C. 1996. *Mathematical Mysteries: The Beauty and Magic of Numbers*. Plenum Press.

Cooper, Necia Grant, ed. 1987. *From Cardinals to Chaos: Reflections on the Life and Legacy of Stanislaw Ulam*. Cambridge University Press, 1989.

Cooper, Roger, April 13, 1991. "The Inside Story," *Spectator*.

Dr. Crypton. December 1985. "Perfect Numbers," *Science Digest*.

Csicsery, George Paul. 1993. *"N Is a Number: A Portrait of Paul Erdős."* Documentary film.

Dalméldico, Amy Dahan. December 1991. "Sophie Germain," *Scientific American*.

Dauben, Joseph Warren. 1979. *Georg Cantor*. Princeton University Press, 1990.

Davis, Philip J., and William G. Chinn. 1969. *3.1416 and All That*. Simon & Schuster.

Davis, Philip J., and Reuben Hersh. 1980. *The Mathematical Experience*. Birkhäuser.

Deacon, Richard. 1985. *The Cambridge Apostles*. Robert Royce.

Derks, Sarah A. September 24, 1996. "Nomadic Genius Lived for Numbers," *The Commercial Appeal* [Memphis].

Devlin, Keith. December 1994. "Mathematician Awarded Nobel Prize," *Focus*, vol. 14, no. 6. Mathematical Association of America.

———. December 1997. "World's Largest Prime," *Focus*, vol. 17, no. 6. Mathematical Association of America.

Dietrich, Bill. August 14, 1996. "It's Prime Time in Seattle for Mathematical Puzzler," *Seattle Times*.

Dostoyevsky, Fyodor. 1880. *The Brothers Karamazov*. Penguin Books, 1993.

Duke, Richard, and Vojtěch Rödl. Summer–Fall 1996. "Paul Erdős Won't Be Coming Anymore," *Newsletter of the SIAM Activity Group on Discrete Mathematics*, vol. 7, no. 1.

Dunham, William. 1990. *Journey Through Genius: The Great Theorems of Mathematics.* John Wiley.

————. 1994. *The Mathematical Universe.* John Wiley.

Eckhardt, Roger. 1987. "Stan Ulam, John von Neumann, and the Monte Carlo Method." In Cooper, Necia Grant, ed., 1987.

*Economist,* October 5, 1996. Obituary, "Paul Erdős."

Eddington, Sir Arthur. 1920. *Space, Time, and Gravitation.* Cambridge University Press.

Erdős, Paul. March 1971. "Child Prodigies." In James Jordan and William A. Webb, eds., *Proceedings of the Washington State University Conference on Number Theory.* Department of Mathematics and Pi Mu Epsilon, Washington State University, Pullman, WA.

————. 1985. "Ulam, the Man and the Mathematician." *Journal of Graph Theory,* vol. 9. John Wiley.

————. 1997. "Some of My Favorite Problems and Results." In Ronald L. Graham and Jaroslav Nešetřil, eds. 1997a.

————, and Ronald L. Graham. July 1975. "On Packing Squares with Equal Squares," *Journal of Combinatorial Theory,* vol. 19, no. 1.

————. 1980. *Old and New Problems and Results in Combinatorial Number Theory.* L'Enseignement Mathématique Université de Genève.

Erdős, Paul, and John L. Selfridge. June 1975. "The Product of Consecutive Integers Is Never a Power," *Illinois Journal of Mathematics,* vol. 19, no. 2.

Fadiman, Clifton. 1957. *Any Number Can Play.* World Publishing Company.

Frisch, Otto. 1979. *What Little I Remember.* Cambridge University Press.

Gamow, George. 1947. *One Two Three...Infinity.* Dover, 1988.

Gardner, Martin. July 1962. "Fiction About Life in Two Dimensions," Mathematical Games, *Scientific American.*

————. March 1964. "The Remarkable Lore of the Prime Numbers," Mathematical Games, *Scientific American.*

————. April 1964. "Various Problems Based on Planar Graphs, or Sets of 'Vertices' Connected by 'Edges,'" Mathematical Games, *Scientific American.*

————. November 1964. "Some Paradoxes and Puzzles Involving Infinite Series and the Concept of Limit," Mathematical Games, *Scientific American.*

————. March 1966. "The Hierarchy of Infinities and the Problems It Spawns," Mathematical Games, *Scientific American*.

————. March 1968. "A Short Treatise on the Useless Elegance of Perfect Numbers and Amicable Paris," Mathematical Games, *Scientific American*.

————. March 1969. "The Multiple Fascinations of the Fibonacci Sequence," Mathematical Games, *Scientific American*.

————. July 1970. "Diophantine Analysis and the Problem of Fermat's Legendary 'Last Theorem,'" Mathematical Games, *Scientific American*.

————. July 1974. "On the Patterns and the Unusual Properties of Figurate Numbers," Mathematical Games, *Scientific American*.

————. June 1977. "The Concept of Negative Numbers and the Difficulty of Grasping It," Mathematical Games, *Scientific American*.

————. October 1978. "Puzzles and Number-Theory Problems Arising from the Curious Fractions of Ancient Egypt," Mathematical Games, *Scientific American*.

————. August 1979. "The Imaginableness of Imaginary Numbers," Mathematical Games, *Scientific American*.

————. October 1979. "Some Packing Problems That Cannot Be Solved by Sitting on the Suitcase," Mathematical Games, *Scientific American*.

Gies, Joseph and Frances. 1969. *Leonardo of Pisa and the New Mathematics of the Middle Ages*. Thomas Y. Crowell.

Gleick, James. October 3, 1993. "Fermat's Theorem," *New York Times Magazine*.

Goffman, Caspar. September 1969. "And What Is Your Erdős Number?" *American Mathematical Monthly*.

Golomb, Michael. May 27, 1977. "Paul Erdős: Addenda," *Science*, vol. 196, no. 4293, p. 938.

————. Fall 1996. "Paul Erdős at Purdue," *Math PUrview*, Department of Mathematics, Purdue University.

Graham, Ronald L. November–December 1964. "A Fibonacci-like Sequence of Composite Numbers," *Mathematics Magazine*, vol. 37, no. 5.

————. 1978. "Combinatorial Scheduling Theory." In Lynn Arthur Steen, ed., *Mathematics Today*. Vintage Books, 1980.

————. March 1978. "The Combinatorial Mathematics of Scheduling," *Scientific American*.

————, and Jaroslav Nešetřil, eds. 1997a. *The Mathematics of Paul Erdős*. Vol. 1. Springer-Verlag.

————, and Jaroslav Nešetřil, eds. 1997b. *The Mathematics of Paul Erdős*. Vol. II. Springer-Verlag.

————, Bruce L. Rothschild, and Joel H. Spencer. 1990. *Ramsey Theory*. John Wiley.

Granville, Andrew, and Ian Katz. June 24, 1993. "The Number's Up for Math's Last Riddle," *The Guardian* [London].

Grossman, Jerrold W. 1997. "Paul Erdős: Master of Collaboration." In Ronald L. Graham and Jaroslav Nešetřil, eds., 1997b.

Guy, Richard K. January 10, 1997. "Erdős, Pál, 1913-03-26 to 1996-09-20." Remarks delivered at the session "A Tribute to Paul Erdős" at the joint meeting of the American Mathematical Society and the Mathematical Association of America, San Diego.

Gyárfás, András. 1997. "Fruit Salad," *Electronic Journal of Combinatorics*, 4, #R8.

Hajnal, András. 1997. "Paul Erdős' Set Theory." In Ronald L. Graham and Jaroslav Nešetřil, eds., 1997b.

Hall, Tord. 1970. Albert Froderberg, trans. *Carl Friedrich Gauss*. MIT Press.

Hardy, G. H. 1940a. *Ramanujan*. Cambridge University Press.

————. 1940b, with a foreword by C. P. Snow. 1967. *A Mathematician's Apology*. Cambridge University Press. 1984.

Hargittai, István. Fall 1997. "A Great Communicator of Mathematics and Other Games: A Conversation with Martin Gardner," *The Mathematical Intelligencer*, vol. 19, no. 4.

Hawkins, David. December 1958. "Mathematical Sieves," *Scientific American*.

Henriksen, Melvin. 1997. "Reminiscences of Paul Erdős (1913–1996)," *http://www.maa.org/features/erdos.html* or *The Humanist Mathematics Network Journal*, 15 (1997).

Henrion, Claudia. 1997. *Women in Mathematics: The Addition of Difference*. Indiana University Press.

Hoensch, Jörg K. 1988. *A History of Modern Hungary 1867–1986*. Longman.

Hoffman, Paul. 1988. *Archimedes' Revenge: The Joys and Perils of Mathematics*. Fawcett Columbine, 1997.

Honsberger, Ross. 1970. *Ingenuity in Mathematics*. Random House.

————. 1973. *Mathematical Gems*. Mathematical Association of America.

————. 1976. *Mathematical Gems II*. Mathematical Association of America.

————. 1979. *Mathematical Plums*. Mathematical Association of America.

————. 1985. *Mathematical Gems III*. Mathematical Association of America.

Horgan, John. March 1997. "Profile: Ronald L. Graham," *Scientific America*.

Ivić, Aleksandar. 1997. "Remembering Paul Erdős." *Nieuw Archief voor Wiskunde*, vol. 15.

Kac, Mark. 1985. *Enigmas of Chance: An Autobiography*. Harper & Row.

Kanigel, Robert. 1991. *The Man Who Knew Infinity: A Life of the Genius Ramanujan*. Charles Scribner's Sons.

Kasner, Edward, and James R. Newman. 1940. *Mathematics and the Imagination*. Simon & Schuster.

Kline, Morris. September 1964. "Geometry," *Scientific American*.

Koestler, Arthur. 1952. *Arrow in the Blue*. Macmillan.

Kolata, Gina. June 24, 1993. "At Last, Shout of 'Eureka!' in Age-Old Math Mystery," *New York Times*.

————. June 29, 1993. "Math Whiz Who Battled 350-Year-Old Problem," *New York Times*.

————. September 24, 1997. "Paul Erdős, 83, a Wayfarer in Math's Vanguard, Is Dead," *New York Times*.

Kramer, Edna E. 1970. *The Nature and Growth of Modern Mathematics*. Princeton University Press, 1982.

Larman, D. G. September 28, 1996. Letters to the editor, *The [London] Times*.

Lemonick, Michael D. July 5, 1993. "*Fini* to Fermat's Last Theorem," *Time*.

Lewis, Harry R., and Christos H. Papadimitriou. January 1978. "The Efficiency of Algorithms," *Scientific American*.

Lukacs, John. 1988. *Budapest 1900*. Grove Press.

Mackenzie, Dana. February 7, 1997. "Homage to an Itinerant Master." *Science*, vol. 275.

Mahoney, Michael Sean. 1973. *The Mathematical Career of Pierre de Fermat*. 2nd ed. Princeton University Press. 1994.

Maor, Eli. 1994. *e: The Story of a Number*. Princeton University Press.
———. 1987. *To Infinity and Beyond: A Cultural History of the Infinite*. Princeton University Press, 1991.

Marshburn, Elizabeth Walker. Summer 1996. "Prove & Conjecture," *Memphis Magazine*, vol. 15, no. 2.

McCagg, William O. 1972. "Jewish Nobles and Geniuses in Modern History," *Eastern European Quarterly*.

Metropolis, N. "The Beginning of the Monte Carlo Method." In Cooper, Necia Grant, ed., 1987.

Milne, E. A. 1948. "Obituary Notices: Godfrey Harold Hardy." *Monthly Notices of the Royal Astronomical Society*, vol. 108.

Motz, Lloyd, and Jefferson Hane Weaver. 1993. *The Story of Mathematics*. Avon Books, 1995.

Nasar, Sylvia. November 13, 1994. "The Lost Years of a Nobel Laureate," *New York Times*. The story of John Nash's schizophrenia.

Nathanson, Melvyn. October 19, 1996. "The Erdős Paradox," preprint based on memorial remarks delivered at the conference "Paul Erdős and His Mathematics," Hungarian Academy of Science, Budapest.

*Nature*. May 22, 1948. "Obituaries: Prof. G. H. Hardy, F.R.S."

Nelson, Carol, and David E. Penny, Carl Pomerance. Spring 1974. "714 and 715," *Journal of Recreational Mathematics*, vol. 7, no. 2.

Newman, James R. 1956. *The World of Mathematics*. Vol. 4. Simon & Schuster.

Oltikar, Sham, and Keith Wayland. January 1983. "Construction of Smith Numbers," *Mathematics Magazine*, vol. 56, no. 1.

Osmond, Warren. October 2, 1996. "Lasting Legacy of Maths Genius," *Campus Review*, vol. 6, no. 38.

Osserman, Robert. 1995. *Poetry of the Universe*. Anchor Books.

Pach, János. Spring 1997. "Two Places at Once: A Remembrance of Paul Erdős," *Mathematical Intelligencer*, vol. 19, no. 2.

Pelikán, József. 1996. "Paul Erdős (1913–1996)," *Mathematical Competitions*, vol. 9, no. 2.

Peterson, Ivars. 1998. *The Jungle of Randomness*. John Wiley.

Petrillo, Lisa. January 11, 1997. "Math Expert Erdős's Career Celebrated at Conference," *San Diego Union-Tribune*.

Pomerance, Carl. January 1998. "Paul Erdős, Number Theorist Extraordinaire," *Notices of the American Mathematical Society*, vol. 45, no. 1.

287

Prakash. M. December 26, 1996. "100 p.c. Pure Talent," *The Hindu* [Madras, India].

Rabson, Alice B. October 15, 1996. "Friendship with Math Genius Was an Honor," Letters to the Editor, *Free Lance-Star* [Fredericksburg, VA].

Raeburn, Paul. June 24, 1993. "Princeton Mathematician Claims Solution to Math's Most Famous Problem," Associated Press.

Regis, Ed. 1987. *Who Got Einstein's Office?* Addison-Wesley.

Reid, Constance. 1964. *From Zero to Infinity*. Thomas Y. Crowell, 1966.

Rhodes, Richard. 1986. *The Making of the Atomic Bomb*. Simon & Schuster.

Ribenboim, Paulo. 1979. *13 Lectures on Fermat's Last Theorem*. Springer-Verlag.

Ribet, Kenneth A., and Brian Hayes. May–April, 1994. "Fermat's Last Theorem and Modern Arithmetic," *American Scientist*, vol. 82.

Roddy, Michael. October 18, 1996. "Mathematicians Mourn Hungary's 'Maths Monk,' " Reuters.

Rota, Gian-Carlo. November 1996. Obituary, *SIAM News*, vol. 29, no. 9. Society for Industrial and Applied Mathematics.

Rucker, Rudy. 1982. *Infinity and the Mind: The Science and Philosophy of the Infinite*. Bantam Books, 1983.

Russell, Bertrand. 1951. *The Autobiography of Bertrand Russell, 1872–1914*. Atlantic Monthly Press, 1967.

———. 1956. *Portraits from Memory and Other Essays*. Allen & Unwin.

———. 1959. *My Philosophical Development*. Allen & Unwin.

Saville, Kirk. March 10, 1995. "Go Figure; Mathematician Paul Erdős Does Not Live Life by the Numbers," *Sun-Sentinel* [Fort Lauderdale].

Sinclair, Lister, host, and Sara Wolch, producer. May 13, 1997. Outtakes from a radio interview with Erdős and Graham. "Math and Aftermath: A Tribute to Donald Coxeter," part 1. Canadian Broadcasting Corporation.

Singh, Simon. 1997. *Fermat's Last Theorem*. Fourth Estate.

———, and Kenneth A. Ribet. November 1977. "Fermat's Last Stand," *Scientific American*, vol. 277, no. 5.

Slatner, David. 1997. *The Joy of π*. Walker.

Smith, Cedric A. B. 1997. "Did Erdős Save Western Civilization?" In Ronald L. Graham and Jaroslav Nešetřil, eds., 1997a.

Snow, C. P. 1966. *Variety of Men*. Scribners.

Spencer, Joel. January 1998. "Uncle Paul," *Notices of the American Mathematical Society*, vol. 45, no. 1.

———, ed. 1973. *Paul Erdős: The Art of Counting, Selected Writings*. MIT Press.

Stewart, Ian. 1987. *From Here to Infinity*. Oxford University Press, 1996.

———. December 17, 1997. "The Formula Man," *New Scientist*.

Stone, A. H. 1997. "Encounters with Paul Erdős." In Ronald L. Graham and Jaroslav Nešetřil, eds., 1997a.

Straus, E. G. 1983. "Paul Erdős at 70," *Combinatorica*, vol. 3 (3–4).

———. "The Elementary Proof of the Prime Number Theorem." Unpublished manuscript.

Sugar, Peter F., ed. 1990. *A History of Hungary*. Indiana University Press.

Szekeres, Gy. 1973. "A Combinatorial Problem in Geometry: Reminiscences." In Spencer, ed. 1973.

Tierney, John. October 1984. "Paul Erdős is in Town. His Brain is Open." *Science 84*, vol. 5, no. 8.

Tőkés, Rudolf L. 1967. *Béla Kun and the Hungarian Soviet Republic*. Frederick A. Praeger.

Ulam, Françoise (without a byline). May 6, 1970. "A Visiting Hungarian Nonagenarian," *Town & Country* [Boulder, CO].

———, ed. 1987. "Conversations with Rota." In Cooper, ed., 1987.

Ulam, Stanislaw M. 1976. *Adventures of a Mathematician*. Charles Scribner's Sons, 1983.

Vázsonyi, Andrew. December 1996. "Paul Erdős," *OR/MS Today*.

———. March 1997. "Paul Erdős, Beloved Math Genius, Leaves Us," *Decision Line*.

Vértesi, Péter. January 1998. "Approximation Theory," *Notices of the American Mathematical Society*, vol. 45, no. 1.

vos Savant, Marilyn. September 9, 1990. "Ask Marilyn," *Parade*.

———. December 2, 1990. "Ask Marilyn," *Parade*.

———. February 17, 1991. "Ask Marilyn," *Parade*.

Wells, David. 1987. *The Penguin Dictionary of Curious and Interesting Numbers*. Penguin Books.

Wilansky, A. 1982. "Smith Numbers," *Two-Year College Mathematics Journal*, vol. 13, no. 21.

Wilf, Herb. January 1997. "A Greeting; and a View of Riemann's Hypothesis," *American Mathematical Monthly*, vol. 94, 1987.

# INDEX